Information Theory
and Evolution 2nd Edition

Information Theory and Evolution 2nd Edition

John Scales Avery
University of Copenhagen, Denmark

World Scientific

NEW JERSEY · LONDON · SINGAPORE · BEIJING · SHANGHAI · HONG KONG · TAIPEI · CHENNAI

Published by

World Scientific Publishing Co. Pte. Ltd.

5 Toh Tuck Link, Singapore 596224

USA office: 27 Warren Street, Suite 401-402, Hackensack, NJ 07601

UK office: 57 Shelton Street, Covent Garden, London WC2H 9HE

British Library Cataloguing-in-Publication Data
A catalogue record for this book is available from the British Library.

Cover photo courtesy of Steve Berardi

INFORMATION THEORY AND EVOLUTION
2nd Edition

ISBN-13 978-981-4401-22-7
ISBN-10 981-4401-22-6
ISBN-13 978-981-4401-23-4 (pbk)
ISBN-10 981-4401-23-4 (pbk)

Printed in Singapore by World Scientific Printers.

Contents

Preface

The aim of this book is to discuss the phenomenon of life, including its origin and evolution (and also including human cultural evolution) against the background of thermodynamics, statistical mechanics and information theory. The second law of thermodynamics states that the entropy (disorder) of the universe always increases. The seeming contradiction between the second law of thermodynamics and the high degree of order and complexity produced by living organisms will be a central theme of the book. This apparent contradiction has its resolution in the information content of the Gibbs free energy which is constantly entering the biosphere from outside sources, as will be discussed in detail in Chapter 4.

The book begins with a sketch of the history of evolutionary thought and research not only during Charles Darwin's lifetime, but also before and after him. Among the pioneers of evolution whose work will be discussed are Aristotle, Condorcet, Linnaeus, Erasmus Darwin, Lamarck, and Lyell. They laid the foundations upon which Charles Darwin built his theory.

After Charles Darwin's death in 1882, the theory of evolution continued to develop and continued to be strengthened by newly discovered facts. Modern molecular biology and DNA technology have allowed us to construct evolutionary family trees in a far more precise way than Darwin and his contemporaries could do on the basis of morphology. Data from comparative sequencing of macromolecules have, on the whole, confirmed the 19th century picture of evolution; but they have also supplied much knowledge which was not available to the early pioneers of evolutionary theory.

Darwin visualized evolution as taking place through natural selection acting on small inheritable variations in the individuals of a species; but we now know that variations can sometimes be sudden and large — through mutations of the type studied by De Vries and Muller, or through the still more drastic mechanism of symbiosis and genetic fusion.

Darwin speculated on the origin of life, but he deliberately omitted discussion of this subject from his publications. However, in the last letter which he is known to have dictated and signed, he wrote: "... the principle of life will hereafter be shown to be a part or consequence of some general law". In our own time, researchers such as A.I. Oparin, Herald Urey, Stanley Miller, Melvin Calvin. Sydney Fox, Leslie Orgel, Carl Sagan, Manfred Eigen, Christian de Duve, Erwin Schrödinger, Claude Shannon, and Stuart Kauffman have begun to uncover this general law.

In the picture that has begun to emerge from the work of these researchers, the earth originally had an atmosphere from which molecular oxygen was almost entirely absent. Energy sources, such as undersea hydrothermal vents, ultraviolet light, volcanism, radioactive decay, lightning flashes, and meteoric impacts, converted the molecules of the earth's primitive ocean and atmosphere into amino acids, nucleotides, and other building blocks of living organisms. Energy-rich molecules, such as H_2S, FeS, H_2, phosphate esters, pyrophosphates, thioesters and HCN were also produced. Since no living organisms were present, and since molecular oxygen was absent from the early atmosphere, the energy-rich molecules were not degraded immediately, and they were present in moderate concentrations in the primitive ocean.

One then visualizes an era of "chemical Darwinism", in which autocatalytic systems competed for the supply of precursors and energy-rich molecules. These autocatalytic systems (i.e., systems of molecules which catalyzed the synthesis of themselves) can be thought of as the precursors of life. They not only "ate" the energy-rich molecules present in the early ocean; they also reproduced; and they competed with each other in a completely Darwinian way, random variations in the direction of greater efficiency being selected and propagated.

An extremely interesting aspect of the picture just discussed is the special role of the energy-rich molecules. They play a special role because the process of molecular Darwinism at first sight seems to be violating the second law of thermodynamics — creating order out of disorder, when according to the second law, disorder ought to be continually increasing. If we reflect further along these lines, all forms of life seem at first sight to be creating order out of disorder, in violation of the second law.

Living organisms are able to do this because they are not closed systems. If we look at the "fine print" of the second law of thermodynamics, it says that the entropy (or disorder) of the universe always increases — and of course it does. Living organisms produce order within themselves and their immediate environments by creating disorder in the universe as a whole. The degradation of food into waste products is, in fact, the process through which life creates local order at the expense of global disorder. Life builds amazing displays of local order; but meanwhile, the disorder of the larger system increases. The larger system includes the sun, the earth, and the cold dust clouds of interstellar space.

In the hypothetical picture of the origin of life presented above, the "food" molecules are degraded by the autocatalysts in the process of order-creating molecular evolution. In Chapter 4 of the present book, we will focus on the entropy relationships in this process. The statistical mechanics of Maxwell, Boltzmann and Gibbs will be compared with information theory, as developed by Claude Shannon and others. It will be shown that Gibbs free energy carries a content of information and that the "thermodynamic information" obtained by the autocatalysts from the free-energy-rich molecules in the primitive ocean was the source of the order which developed during the process of chemical evolution.

Today, the earth's greatest source of thermodynamic information is the flood of free energy which reaches us in the form of photons from the sun. In Chapter 4, a quantitative relationship will be derived connecting the energy of an absorbed photon and its information content. Readers who wish to skip the mathematics in Chapter 4 may do so without losing the thread of the argument, provided that they are willing to accept on faith the main result of the derivations — the fact that Gibbs free energy contains the thermodynamic form of information.

It seems probable that thermodynamic information derived from free energy was the driving force behind the origin of life. It is today the driving force behind all forms of life — behind the local order which life is able to produce. This is the "general law" which Darwin guessed might someday be shown to underlie the principle of life. All of the information contained in the complex, beautiful, and statistically unlikely structures which are so characteristic of living organisms can be seen as having been distilled from the enormous flood of thermodynamic information which reaches the earth in the form of sunlight.

Where do humans fit into this picture? Like all other forms of life on earth, humans pass information from one generation to the next, coded into the base sequences of their DNA. However, humans have developed a second, highly effective mode of information transmission — language and culture.

Although language and culture are not unique to our species, the extent to which they are developed is unique on earth. Thus humans are distinguished from other species by having two modes of evolutionary change — genetic evolution, symbolized by the long information-containing DNA molecule, and cultural evolution, which might be symbolized by a book or a computer diskette.

If we compare these two modes of evolution, we can see that genetic evolution is very slow, while cultural evolution is extremely rapid — and accelerating. The human genome has changed very little during the last 40,000 years; but during this period, cultural evolution has altered our way of life beyond recognition. Therefore human nature, formed to fit the way of life of our hunter-gatherer ancestors, is not entirely appropriate for our present way of life. For example, human nature seems to contain an element of what might be called "tribalism", which does not fit well with the modern world's instantaneous communications and increasing interdependence.

Not only does the genetic evolution of humans lag behind their cultural evolution, but also cultural evolution itself has a rapidly-moving component and a slowly-moving component which lags behind, creating tensions. As we enter the 21st century, technology is developing with phenomenal speed, while social and political institutions change far more slowly. The disharmony thus created requires study and thought if human society is not to be shaken to pieces by the rapidity of scientific progress.

Interestingly, information technology and biotechnology, the two most rapidly developing fields, are becoming increasingly linked, each finding inspiration in the other. Biologists have studied the mechanism of self-assembly of supramolecular structures such as cell membranes, viruses, chloroplasts and mitochondria. Researchers in the field of nanoscience are now attempting to use this principle of supramolecular organization, observed in biology, to reach a new degree of miniaturization for the switches and memory devices of information technology. Simulated evolution, modelled after biological evolution, has been used to develop new and unorthodox computer hardware and software. Meanwhile, computers and automation are becoming more and more essential to biotechnology; and in fact many universities now have departments devoted to bioinformatics. Chapter 7 will trace the history of information technology, while Chapter 8 will discuss the ways in which it is merging with biotechnology.

The final chapter of the book looks at the future of the new field, bioinformation technology, attempting to predict what it will achieve during the new century, and discussing how these achievements will affect society.

Acknowledgements

I am extremely grateful to Professor Lawrence S. Lerner, Dr. Cindy Lerner, Professor Dudley R. Herschbach, Dr. Anita Goel, Dr. Luis Emilio Bruni, and to my son, Dr. James Avery, for their generous advice and help. The contributions of the Lerners have been especially great. Their careful reading of the entire manuscript and their many detailed suggestions have been invaluable. My son James read several versions of Chapter 4, and his advice was a great help in writing the final version of this central chapter. He also helped me to improve Chapters 7 and 8. A few sections of this book are based on notes for a course on Science and Society which I have taught at the University of Copenhagen since 1989. Other sections draw inspiration from a course on Statistical Mechanics, seen from the standpoint of information theory, which I helped to teach some years earlier. The Statistical Mechanics course was planned by Associate Professor Knud Andersen. I am also grateful to Professor Albert Szent-Györgyi, one of the most important pioneers of Bioenergetics, for many inspiring conversations during the years when I worked in his laboratory at Woods Hole. Finally I would like to thank Mr. Henning Vibæk, who drew the figures.

Introduction to the second edition

I am grateful to the World Scientific Publishing Company for asking me to prepare a second edition of this book. This has allowed me to make some additions which I hope readers will enjoy:

In Chapter 3, I have added a little about the work of Prof. Jack Sjostak and his laboratory at Harvard Medical School, where work is underway aimed at building a synthetic cellular system that undergoes Darwinian evolution.

Another addition to Chapter 3 is a discussion of the picture of the origin of life that was proposed by Michael J. Russell and Alan J. Hall in 1997. In this picture life emerged approximately 4 billion years ago at sites on the deep ocean floor where alkaline, sulphide-bearing submarine seepage waters interfaced with colder, more oxidized, more acid waters, bearing Fe^{2+} ions. According to Russell and Hall, the geochemistry occurring at this interface and continuing over long periods of time, generated organic compounds that were the precursors of life. The scenario proposed by Russell and Hall has some similarities to the prior proposal of Wächterhäuser (discussed in the first edition). However, there are also differences.

A section added to Chapter 4 compares the characteristics of thermodynamic information with those of cybernetic (or semiotic) information. As is pointed out in this section, the evolutionary process consists in making many copies of a molecule or a larger system, and these undergo random mutations. After further copying, natural selection preserves those mutations that are favorable. It is thermodynamic information that drives the copying process, while the selected favorable mutations may be said to contain cybernetic information.

Much more extensive additions have been made to Chapter 6, incorporating the exciting new DNA research that has thrown much light on human prehistory. From this research one can deduce the story of how a surprisingly small group of modern humans left Africa roughly 60,000 years before the present and populated the remainder of the earth. Linguistic studies have also thrown light on these events, and they broadly support the picture deduced from the studies of Y-chromosomal and mitochondrial DNA.

I have attempted to bring Chapter 7 up to date. However, because of limitations of space I have not tried to update Chapter 8 ("Bioinformation Technology"). Research in this field thunders ahead with enormous speed. Interested readers are referred to the recent publications of MIT's Department of Brain and Cognative Sciences.

An extensive Appendix C has been added on "Entropy and Economics". It discusses the aspects of human cultural evolution that involve large amounts of matter and energy. Alfred Lotka (1820–1949), coined the terms "exosomatic evolution" and "endosomatic evolution", which have the same meaning as "cultural evolution" and "genetic evolution", except that Lotka's exosomatic evolution also covers not only the growth of knowledge but also the development and proliferation of human artifacts. This point of view is also found in the pioneering work of Nicholas Georgescu-Roegen (1906–1994), who compared the human economy to the digestive system of human society and who introduced the concept of entropy into economic theory. His work, and that of his distinguished student Herman Daly, (1938–), are discussed in Appendix C, as well as the relationship of these ideas to Bioeconomics.

I have also corrected a notorious typographical error in Appendix B of the first edition, where the date of birth of Jakob von Uexküll is given as 1894, implying that he was awarded an honorary degree by the University of Heidelberg at the age of 13. Everyone agrees that von Uexküll was a very brilliant man, but still 13 seems a bit young for such a high honor. His correct date of birth, 1864, is now in place.

Chapter 1

PIONEERS OF EVOLUTIONARY THOUGHT

Aristotle

Aristotle was born in 381 B.C., the son of the court physician of the king of Macedon, and at the age of seventeen he went to Athens to study. He joined Plato's Academy and worked there for twenty years until Plato died. Aristotle then left the Academy, saying that he disapproved of the emphasis on mathematics and theory and the decline of natural science. After serving as tutor for Alexander of Macedon, he founded a school of his own called the Lyceum. At the Lyceum, he built up a collection of manuscripts which resembled the library of a modern university.

Aristotle was a very great organizer of knowledge, and his writings almost form a one-man encyclopedia. His best work was in biology, where he studied and classified more than five hundred animal species, many of which he also dissected. In Aristotle's classification of living things, he shows an awareness of the interrelatedness of species. This interrelatedness was much later used by Darwin as evidence for the theory of evolution. One cannot really say that Aristotle developed a theory of evolution, but he was groping towards the idea. In his history of animals, he writes:

"Nature proceeds little by little from lifeless things to animal life, so that it is impossible to determine either the exact line of demarcation, or—on which side of the line an intermediate form should lie. Thus, next after lifeless things in the upward scale comes the plant. Of plants, one will differ from another as to its apparent amount of vitality. In a word, the whole plant kingdom, whilst devoid of life as compared with the animal, is yet endowed with life as compared with other corporeal entities. Indeed, there is observed in plants a continuous scale of ascent towards the animal".

Aristotle's classification of living things, starting at the bottom of the scale and going upward, is as follows: Inanimate matter, lower plants and

1

sponges, higher plants, jellyfish, zoophytes and ascidians, molluscs, insects, jointed shellfish, octopuses and squids, fish and reptiles, whales, land mammals and man. The acuteness of Aristotle's observation and analysis can be seen from the fact that he classified whales and dolphins as mammals (where they belong) rather than as fish (where they superficially seem to belong, and where many ancient writers placed them).

Among Aristotle's biological writings, there appears a statement that clearly foreshadows the principle of natural selection, later independently discovered by Darwin and Wallace and fully developed by Darwin. Aristotle wrote: "Wheresoever, therefore... all parts of one whole happened like as if they were made for something, these were preserved, having been appropriately constituted by an internal spontaneity; and wheresoever things were not thus constituted, they perished, and still perish".

One of Aristotle's important biological studies was his embryological investigation of the developing chick. Ever since his time, the chick has been the classical object for embryological studies. He also studied the four-chambered stomach of the ruminants and the detailed anatomy of the mammalian reproductive system. He used diagrams to illustrate complex anatomical relationships — an important innovation in teaching technique.

Averröes

During the Middle Ages, Aristotle's evolutionary ideas were revived and extended in the writings of the Islamic philosopher Averröes[1], who lived in Spain from 1126 to 1198. His writings had a great influence on western thought. Averroes shocked both his Moslem and his Christian readers by his thoughtful commentaries on the works of Aristotle, in which he maintained that the world was not created at a definite instant, but that it instead evolved over a long period of time, and is still evolving.

Like Aristotle, Averröes seems to have been groping towards the ideas of evolution which were later developed in geology by Lyell and in biology by Darwin and Wallace. Much of the scholastic philosophy written at the University of Paris during the 13th century was aimed at refuting the doctrines of Averroes; but nevertheless, his ideas survived and helped to shape the modern picture of the world.

---

[1] Abul Walid Mahommed Ibn Achmed, Ibn Mahommed Ibn Rosched.

The mystery of fossils

During the lifetime of Leonardo da Vinci (1452–1519) the existence of fossil shells in the rocks of high mountain ranges was recognized and discussed. "...the shells in Lombardy are at four levels", Leonardo wrote, "and thus it is everywhere, having been made at various times... The stratified stones of the mountains are all layers of clay, deposited one above the other by the various floods of the rivers". Leonardo had no patience with the explanation given by some of his contemporaries, that the shells had been carried to mountain tops by the deluge described in the Bible. "If the shells had been carried by the muddy waters of the deluge", he wrote, "they would have been mixed up, and separated from each other amidst the mud, and not in regular steps and layers". Nor did Leonardo agree with the opinion that the shells somehow grew within the rocks: "Such an opinion cannot exist in a brain of much reason", he wrote, "because here are the years of their growth, numbered on their shells, and there are large and small ones to be seen, which could not have grown without food, and could not have fed without motion...and here they could not move".

Leonardo believed that the fossil shells were once part of living organisms, that they were buried in strata under water, and much later lifted to the tops of mountains by geological upheavals. However his acute observations had little influence on the opinions of his contemporaries because they appear among the 4000 or so pages of notes which he wrote for himself but never published.

It was left to the Danish scientist Niels Stensen (1638–1686) (usually known by his Latinized name, Steno) to independently rediscover and popularize the correct interpretation of fossils and of rock strata. Steno, who had studied medicine at the University of Leiden, was working in Florence, where his anatomical studies attracted the attention of the Grand Duke of Tuscany, Ferdenand II. When an enormous shark was caught by local fishermen, the Duke ordered that its head be brought to Steno for dissection. The Danish anatomist was struck by shape of the shark's teeth, which reminded him of certain curiously shaped stones called glossopetrae that were sometimes found embedded in larger rocks. Steno concluded that the similarity of form was not just a coincidence, and that the glossopetrae were in fact the teeth of once-living sharks which had become embedded in the muddy sediments at the bottom of the sea and gradually changed to stone. Steno used the corpuscular theory of matter, a forerunner of atomic theory, to explain how the composition of the fossils could have changed while their form remained constant. Steno also formulated a law of strata, which states that in the deposition of layers of sediment, later converted to rock, the oldest layers are at the bottom.

In England, the brilliant and versatile experimental scientist Robert Hooke (1635–1703) added to Steno's correct interpretation of fossils by noticing that some fossil species are not represented by any living counterparts. He concluded that "there have been many other Species of Creatures in former Ages, of which we can find none at present; and that 'tis not unlikely also but that there may be divers new kinds now, which have not been from the beginning".

Similar observations were made by the French naturalist, Georges-Louis Leclerc, Comte de Buffon (1707–1788), who wrote: "We have monuments taken from the bosom of the Earth, especially from the bottom of coal and slate mines, that demonstrate to us that some of the fish and plants that these materials contain do not belong to species currently existing". Buffon's position as keeper of the Jardin du Roi, the French botanical gardens, allowed him time for writing, and while holding this post he produced a 44-volume encyclopedia of natural history. In this enormous, clearly written, and popular work, Buffon challenged the theological doctrines which maintained that all species were created independently, simultaneously and miraculously, 6000 years ago. As evidence that species change, Buffon pointed to vestigial organs, such as the lateral toes of the pig, which may have had a use for the ancestors of the pig. He thought that the donkey might be a degenerate relative of the horse. Buffon believed the earth to be much older than the 6000 years allowed by the Bible, but his estimate, 75,000 years, greatly underestimated the true age of the earth.

The great Scottish geologist James Hutton (1726–1797) had a far more realistic picture of the true age of the earth. Hutton observed that some rocks seemed to have been produced by the compression of sediments laid down under water, while other rocks appeared to have hardened after previous melting. Thus he classified rocks as being either igneous or else sedimentary. He believed the features of the earth to have been produced by the slow action of wind, rain, earthquakes and other forces which can be observed today, and that these forces never acted with greater speed than they do now. This implied that the earth must be immensely old, and Hutton thought its age to be almost infinite. He believed that the forces which turned sea beds into mountain ranges drew their energy from the heat of the earth's molten core. Together with Steno, Hutton is considered to be one of the fathers of modern geology. His uniformitarian principles, and his belief in the great age of the earth were later given wide circulation by Charles Darwin's friend and mentor, Sir Charles Lyell (1797–1875), and they paved the way for Darwin's application of uniformitarianism to biology. At the time of his death, Hutton was working on a theory of biological evolution through natural selection, but his manuscripts on this subject remained unknown until 1946.

Condorcet

Further contributions to the idea of evolution were made by the French mathematician and social philosopher Marie-Jean-Antoine-Nicolas Caritat, Marquis de Condorcet, who was born in 1743. In 1765, when he was barely 22 years old, Condorcet presented an *Essay on the Integral Calculus* to the Academy of Sciences in Paris. The year 1785 saw the publication of Condorcet's highly original mathematical work, *Essai sur l'application de l'analyse à la probabilité des decisions rendues à la pluralité des voix*[2], in which he pioneered the application of the theory of probability to the social sciences. A later, much enlarged, edition of this book extended the applications to games of chance.

Condorcet had also been occupied, since early childhood, with the idea of human perfectibility. He was convinced that the primary duty of every person is to contribute as much as possible to the development of mankind, and that by making such a contribution, one can also achieve the greatest possible personal happiness. When the French Revolution broke out in 1789, he saw it as an unprecedented opportunity to do his part in the cause of progress; and he entered the arena wholeheartedly, eventually becoming President of the Legislative Assembly, and one of the chief authors of the proclamation which declared France to be a republic. Unfortunately, Condorcet became a bitter enemy of the powerful revolutionary politician, Robespierre, and he was forced to go into hiding.

Although Robespierre's agents had been unable to arrest him, Condorcet was sentenced to the guillotine in absentia. He knew that in all probability he had only a few weeks or months to live; and he began to write his last thoughts, racing against time. Condorcet returned to a project which he had begun in 1772, a history of the progress of human culture, stretching from the remote past to the distant future. Guessing that he would not have time to complete the full-scale work he had once planned,

---

[2] *Essay on the Application of Analysis to the Probability of Decisions Taken According to a Plurality of Votes.*

he began a sketch or outline: *Esquisse d'un tableau historique des progrès de l'esprit humain*[3].

In his *Esquisse*, Condorcet enthusiastically endorsed the idea of infinite human perfectibility which was current among the philosophers of the 18th century; and he anticipated many of the evolutionary ideas which Charles Darwin later put forward. He compared humans with animals, and found many common traits. According to Condorcet, animals are able to think, and even to think rationally, although their thoughts are extremely simple compared with those of humans. Condorcet believed that humans historically began their existence on the same level as animals and gradually developed to their present state. Since this evolution took place historically, he reasoned, it is probable, or even inevitable, that a similar evolution in the future will bring mankind to a level of physical, mental and moral development which will be as superior to our own present state as we are now superior to animals.

At the beginning of his manuscript, Condorcet stated his belief "that nature has set no bounds on the improvement of human facilities; that the perfectibility of man is really indefinite; and that its progress is henceforth independent of any power to arrest it, and has no limit except the duration of the globe upon which nature has placed us". He stated also that "the moral goodness of man is a necessary result of his organism; and it is, like all his other facilities, capable of indefinite improvement".

like the other scientists and philosophers of his period, Condorcet accepted the Newtonian idea of an orderly cosmos ruled by natural laws to which there are no exceptions. He asserted that the same natural laws must govern human evolution, since humans are also part of nature. Again and again, Condorcet stressed the fundamental similarity between humans and animals; and he regarded all living things as belonging to the same great family. (It is perhaps this insight which made Condorcet so sensitive to the feelings of animals that he even avoided killing insects.) To explain the present differences between humans and animals, Condorcet maintained, we need only imagine gradual changes, continuing over an extremely long period of time. These long-continued small changes have very slowly improved human mental abilities and social organization, so that now, at the end of an immense interval of time, large differences have appeared between ourselves and lower forms of life.

Condorcet regarded the family as the original social unit; and in *Es-*

---

[3] *Sketch of an Historical Picture of the Progress of the Human Spirit.*

quisse he called attention to the unusually long period of dependency which characterizes the growth and education of human offspring. This prolonged childhood is unique among living beings. It is needed for the high level of mental development of the human species; but it requires a stable family structure to protect the young during their long upbringing. Thus, according to Condorcet, biological evolution brought into existence a moral precept, the sanctity of the family.

Similarly, Condorcet wrote, larger associations of humans would have been impossible without some degree of altruism and sensitivity to the suffering of others incorporated into human behavior, either as instincts or as moral precepts or both; and thus the evolution of organized society entailed the development of sensibility and morality. Unlike Rousseau, Condorcet did not regard humans in organized civilizations as degraded and corrupt compared to "natural" man; instead he saw civilized humans as more developed than their primitive ancestors.

Believing that ignorance and error are responsible for vice, Condorcet discussed what he believed to be the main mistakes of civilization. Among these he named hereditary transmission of power, inequality between men and women, religious bigotry, disease, war, slavery, economic inequality, and the division of humanity into mutually exclusive linguistic groups. Regarding disease, Condorcet predicted that the progress of medical science would ultimately abolish it. Also, he maintained that since perfectibility (i.e., evolution) operates throughout the biological world, there is no reason why mankind's physical structure might not gradually improve, with the result that human life in the remote future could be greatly prolonged.

Condorcet believed that the intellectual and moral facilities of man are capable of continuous and steady improvement; and he thought that one of the most important results of this improvement would be the abolition of war. As humans become enlightened in the future (he believed) they will recognize war as an atrocious and unnecessary cause of suffering; and as popular governments replace hereditary ones, wars fought for dynastic reasons will disappear. Next to vanish will be wars fought because of conflicting commercial interests. Finally, the introduction of a universal language throughout the world and the construction of perpetual confederations between nations will eliminate, Condorcet predicted, wars based on ethnic rivalries.

With better laws, social and financial inequalities would tend to become leveled. To make the social conditions of the working class more equal to those of the wealthy, Condorcet advocated a system of insurance (either private or governmental) where the savings of workers would be used to provide pensions and to care for widows and orphans. Also, since social inequality is related to inequality of education, Condorcet advocated a system of universal public education supported by the state.

At the end of his *Esquisse*, Condorcet wrote that any person who has contributed to the best of his ability to the progress of mankind becomes immune to personal disaster and suffering. He knows that human progress is inevitable, and can take comfort and courage from his inner picture of the epic march of mankind, through history, towards a better future. Eventually Condorcet's hiding-place was discovered. He fled in disguise, but was arrested after a few days; and he died soon afterwards in his prison cell. After Condorcet's death the currents of revolutionary politics shifted direction. Robespierre, the leader of the Terror, was himself soon arrested. The execution of Robespierre took place on July 25, 1794, only a few months after the death of Condorcet.

Condorcet's *Esquisse d'un tableau historique des progrès de l'esprit h main* was published posthumously in 1795. In the post-Thermidor reconstruction, the Convention voted funds to have it printed in a large edition and distributed throughout France, thus adopting the *Esquisse* as its official manifesto. This small but prophetic book is the one for which Condorcet is now chiefly remembered. It was destined to establish the form in which the eighteenth-century idea of progress was incorporated into Western thought, and it provoked Robert Malthus to write *An Essay on the Principle of Population*. Condorcet's ideas are important because he considered the genetic evolution of plants and animals and human cultural evolution to be two parts of a single process.

Linnaeus

Meanwhile, during the 17th and 18th centuries, naturalists had been gathering information on thousands of species of plants and animals. This huge, undigested heap of information was put into some order by the great Swedish naturalist, Carl von Linné (1707–1778), who is usually called by his Latin name, Carolus Linnaeus.

Linnaeus was the son of a Swedish pastor. Even as a young boy, he was fond of botany, and after medical studies at Lund, he became a lecturer in botany at the University of Uppsala, near Stockholm. In 1732, the 25-year-old Linnaeus was asked by his university to visit Lapland to study the plants in that remote northern region of Sweden.

Linnaeus travelled four thousand six hundred miles in Lapland, and he discovered more than a hundred new plant species. In 1735, he published his famous book, *Systema Naturae*, in which he introduced a method for the classification of all living things.

Linnaeus not only arranged closely related species into genera, but he also grouped related genera into classes, and related classes into orders. (Later the French naturalist Cuvier (1769–1832) extended this system by grouping related orders into phyla.) Linnaeus introduced the binomial nomenclature, still used today, in which each plant or animal is given a name whose second part denotes the species while the first part denotes the genus.

Although he started a line of study which led inevitably to the theory of evolution, Linnaeus himself believed that species are immutable. He adhered to the then-conventional view that each species had been independently and miraculously created six thousand years ago, as described in the Book of Genesis.

Linnaeus did not attempt to explain why the different species within a genus resemble each other, nor why certain genera are related and can be grouped into classes, etc. It was not until a century later that these resemblances were understood as true family likenesses, so that the resemblance between a cat and a lion came to be understood in terms of their descent from a common ancestor[4].

Erasmus Darwin

Among the ardent admirers of Linnaeus was the brilliant physician-poet, Erasmus Darwin (1731–1802), who was considered by Coleridge to have "...a greater range of knowledge than any other man in Europe". He was also the best English physician of his time, and George III wished to have him as his personal doctor. However, Darwin preferred to live in the north of England rather than in London, and he refused the position.

In 1789, Erasmus Darwin published a book called *The Botanic Garden or The Loves of the Plants*. It was a book of botany written in verse, and in the preface Darwin stated that his purpose was "...to inlist imagination under the banner of science..." and to call the reader's attention to "the

---

[4] Linnaeus was to Darwin what Kepler was to Newton. Kepler accurately described the motions of the solar system, but it remained for Newton to explain the underlying dynamical mechanism. Similarly, Linnaeus set forth a descriptive "family tree" of living things, but Darwin discovered the dynamic mechanism that underlies the observations.

immortal works of the celebrated Swedish naturalist, Linnaeus". This book was immensely popular at the time when it was written, but it was later satirized by Pitt's Foreign Minister, Canning, whose book *The Loves of the Triangles* ridiculed Darwin's poetic style.

In 1796, Erasmus Darwin published another book, entitled *Zoonomia*, in which he proposed a theory of evolution similar to that which his grandson, Charles Darwin, was later to make famous. "...When we think over the great changes introduced into various animals", Darwin wrote, "as in horses, which we have exercised for different purposes of strength and swiftness, carrying burthens or in running races; or in dogs, which have been cultivated for strength and courage, as the bull-dog; or for acuteness of his sense of smell, as in the hound and spaniel; or for the swiftness of his feet, as the greyhound; or for his swimming in the water, or for drawing snow-sledges, as the rough-haired dogs of the north... and add to these the great change of shape and color which we daily see produced in smaller animals from our domestication of them, as rabbits or pigeons;... when we revolve in our minds the great similarity of structure which obtains in all the warm-blooded animals, as well as quadrupeds, birds and amphibious animals, as in mankind, from the mouse and the bat to the elephant and whale; we are led to conclude that they have alike been produced from a similar living filament."

"Would it be too bold", Erasmus Darwin asked, "to imagine that in the great length of time since the earth began to exist, perhaps millions of ages before the commencement of the history of mankind — would it be to bold to imagine that all warm-blooded animals have arisen from one living filament?"

Lamarck

In France, Jean Baptiste Pierre Antoine de Monet, Chevalier de Lamarck (1744–1829), contributed importantly to the development of evolutionary ideas. After a period in the French army, from which he was forced to retire because of illness, Lamarck became botanist to the king, and later Professor of Invertebrate Zoology at the Museum of Natural History in Paris. Lamarck deserves to be called the father of invertebrate zoology. Linnaeus had exhausted his energy on the vertebrates, and he had left the invertebrates in disorder. Their classification is largely due to Lamarck: He differentiated the eight-legged arachnids, such as spiders and scorpions, from six-legged insects; he established the category of crustaceans for crabs, lobsters etc.; and he introduced the category of echinoderms for starfish, sea-urchins etc. Between 1785 and 1822, Lamarck published seven huge volumes of a treatise entitled *Natural History of Invertebrates*. However, it is for his book *Zoological Philosophy*, published in 1809, that the Chevalier de Lamarck is chiefly remembered today.

In his *Zoological Philosophy*, Lamarck stated his belief that the species within a genus owe their similarity to descent from a common ancestor. He was the first prominent biologist since the age of Aristotle to believe that species are not immutable but that they have changed during the long history of the earth.

Although Lamarck deserves much credit as a pioneer of evolutionary thought, he was seriously wrong about the mechanism of change. For example, Lamarck believed that the long neck of the giraffe evolved because each giraffe stretched its neck slightly in an effort to reach the leaves on high trees. He believed that these slightly-stretched necks could be inherited, and thus, in this way, over many generations, the necks of giraffes had grown longer and longer. Although Lamarck was right in his general picture of evolution, he was mistaken in the detailed mechanism which he proposed, since later experiments proved conclusively that, in general, acquired characteristics cannot be inherited. (One must say "in general", because in the case of symbiosis and genetic fusion, acquired characteristics are inherited. Plasmids containing genetic material are also frequently exchanged between bacteria. Furthermore, in human cultural evolution, innovations can be passed on to future generations. We will discuss these Lamarckian mechanisms of evolution in later chapters.)

The debates between Cuvier and Geoffroy St. Hilaire

In 1830, a year after the death of Lamarck, a famous series of debates took place between Georges Leopold Dagobert, Baron Cuvier (1769–1832) and Etienne Geoffroy St. Hilaire (1772–1844). The two men, both professors at the Musee National d'Histoire Naturelle in Paris, were close friends and scientific collaborators. However, they differed in their opinions, especially on the question of whether the form of an animal's parts led to their function, or whether the reverse was true. Cuvier almost singlehandedly founded the discipline of vertebrate paleontology, and he firmly established the fact that extinctions have taken place. However, he did not believe in evolution. In 1828, Cuvier wrote: "If there are resemblances between the organs of fishes and those of other vertebrate classes, it is only insofar as there are resemblances between their functions". In other words, function produces form. Cuvier denied that similarity of form implied descent from a common ancestor.

St. Hilaire, on the other hand, considered all vertebrates to be modifications of a single archetype. He maintained that similar vestigial organs and similarities in embryonic development implied descent from a common ancestor. He was especially interested in homologies, that is, cases where similar structures in two different organisms are used for two different purposes. In 1829, St. Hilaire wrote: "Animals have no habits but those that result from the structure of their organs: if the latter varies, there vary in the same manner all their springs of action, all their facilities, and all their actions".

The opposing viewpoints of the two men led to a famous series of eight public debates, which took place from February to April, 1830. Although Cuvier was thought by most observers to have won the debates, St. Hilaire's belief in evolution continued, as did the friendship between the two naturalists. In 1832 St. Hilaire partially anticipated Darwin's theory of evolution through natural selection: "The external world is all-powerful in alteration of the form of organized bodies...", he wrote, "These [modifications] are inherited, and they influence all the rest of the organization of the animal, because if these modifications lead to injurious effects, the animals which exhibit them perish and are replaced by others of a somewhat different form, a form changed so as to be adapted to the new environment".

Suggestions for further reading

(1) P.J. Bowler, *Evolution: The History of an Idea*, University of California Press, (1989).
(2) D.J. Putuyma, *Evolutionary Biology*, Sinauer Associates, Sunderland Mass., (1986).
(3) B. Glass, 0. Temkin, and W.L. Strauss, eds., *Forerunners of Darwin: 1745-1859*, Johns Hopkins Press, Baltimore, (1959).
(4) R. Milner, *The Encyclopedia of Evolution*, an Owl Book, Henry Holt and Company, New York, (1990).
(5) T.A. Appel, *The Cuvier-Geoffroy Debate: French Biology in the Decades before Darwin*, Oxford University Press, (1987).
(6) P.J. Bowler, *Fossils and Progress: Paleontology and the Idea of Progressive Evolution in the Nineteenth Century*, Science History Publications, New York, (1976).

(7) H. Torrens, *Presidential Address: Mary Anning (1799-1847) of Lyme; 'the greatest fossilist the world ever knew'*, British Journal of the History of Science, **28**, 257-284, (1995).

(8) P. Corsi, *The Age of Lamarck: Evolutionary Theories in France, 1790-1834*, University of California Press, Berkeley, (1988).

(9) C.C. Gillispie, *Genesis and Geology: A Study in the Relations of Scientific Thought, Natural Theology and Social Opinion in Great Britain, 1790-1850*, Harvard University Press, Cambridge Mass., (1951).

(10) M. McNeil, *Under the Banner of Science: Erasmus Darwin and his Age*, Manchester University Press, Manchester, (1987).

(11) L.G. Wilson, *Sir Charles Lyell's Scientific Journals on the Species Question*, Yale University Press, New Haven, (1970).

(12) M. 'Espinasse, *Robert Hooke, 2nd ed.*, U. of California Press, (1962).

(13) M.J.S. Rudwick, *The Meaning of Fossils: Episodes in the History of Paleontology, 2nd ed.*, University of Chicago Press, (1985).

(14) A.B. Adams, *Eternal Quest: The Story of the Great Naturalists*, G.P. Putnam's Sons, New York, (1969).

(15) A.S. Packard, *Lamarck, the Founder of Evolution: His Life and Work*, Longmans, Green, and Co., New York, (1901).

(16) C. Darwin, *An historical sketch of the progress of opinion on the Origin of Species, previously to the publication of this work*, Appended to third and later editions of *On the Origin of Species*, (1861).

(17) L. Eiseley, *Darwin's Century: Evolution and the Men who Discovered It*, Doubleday, New York, (1958).

(18) H.F. Osborne, *From the Greeks to Darwin: The Development of the Evolution Idea Through Twenty-Four Centuries*, Charles Scribner and Sons, New York, (1929).

Chapter 2

CHARLES DARWIN'S LIFE AND WORK

Family background and early life

It was Erasmus Darwin's grandson Charles (1809–1882) who finally worked out a detailed and correct theory of evolution and supported it by a massive weight of evidence.

As a boy, Charles Darwin was passionately fond of hunting and collecting beetles, but he was a mediocre student. His father once said to him in exasperation: "You care for nothing but shooting, dogs and rat-catching; and you will be a disgrace to yourself, and to all your family!"

Darwin's father, a wealthy physician, sent him to Edinburgh University to study medicine; but Charles did not enjoy his studies there. "Dr. Duncan's lectures on Materia Medica at 8 o'clock on a winter's morning are something fearful to remember", he wrote later. "I also attended the operating theatre in the hospital at Edinburgh and saw two very bad operations, one on a child, but I rushed away before they were completed. Nor did I ever attend again, for hardly any inducement would have been strong enough to make me do so; this being long before the blessed days of chloroform. The two cases fairly haunted me for many a long year".

The time at Edinburgh was not entirely wasted, however, because several of Darwin's friends at the university were natural philosophers[1], and contact with them helped to develop his interest in natural history. One of the most important of these scientific friends was Dr. R.E. Grant, an expert on marine invertebrate zoology with whom Darwin often collected small sea slugs in the cold waters of the Firth near Edinburgh. On one of these expeditions, Grant suddenly began to praise the evolutionary views of Lamarck, while Darwin listened in silent astonishment. Charles Darwin

---

[1] Today we would call them scientists.

had previously read his own grandfather's book *Zoonomia* and had greatly admired it; but after a few years he had read it again in a more critical spirit; and after the second reading he had decided that *Zoonomia* was too speculative and contained too few facts. Grant's praise of Lamarck may have helped Darwin to become, later in his life, an advocate of evolution in a different form.

Darwin's father finally gave up the idea of making him into a doctor, and sent him instead to Cambridge to study for the clergy. At Cambridge, Darwin made many friends because of his unfailing good nature, enthusiasm and kindness. A friend from university days remembers that "at breakfast, wine or supper parties he was ever one of the most cheerful, the most popular and the most welcome... He was the most genial, warmhearted, generous and affectionate of friends".

Darwin's best friend during his last two years at Cambridge was the Reverend John Stevens Henslow, Professor of Botany. Darwin was often invited to Henslow's family dinner; and on most days he accompanied the professor on long walks, so that he became known as "the man who walks with Henslow". This friendship did much to develop Darwin's taste for natural history. Henslow's knowledge of botany, zoology and geology was vast; and he transmitted much of it to his enthusiastic young student during their long walks through the beautiful countryside near to the university. At Cambridge Darwin collected beetles; and the hobby became almost a passion for him. "One day, on tearing off some old bark", he wrote later, "I saw two rare beetles, and seized one in each hand. Then I saw a third kind, which I could not bear to lose, so I popped the one held in my right hand into my mouth. Alas! It ejected some intensely acrid fluid which burnt my tongue, so that I was forced to spit the beetle out, which was lost, as was the third one".

During his last year at Cambridge, Darwin read Alexander von Humboldt's famous *Personal Narrative of Travels to the Equinoctial Regions of South America During the Years 1799–1804*, a book which awakened in him "a burning zeal to add even the most humble contribution to the noble structure of Natural Science". Darwin longed to visit the glorious tropical forests described so vividly by von Humboldt.

Henslow persuaded Darwin to begin to study geology; and during the spring of 1831, Darwin joined the Professor of Geology, Adam Sedgwick, on an expedition to study the ancient rock formations in Wales. This expedition made Darwin realize that "science consists in grouping facts in such a way that general laws or conclusions may be drawn from them".

When Darwin returned from Wales, he found a letter from Professor George Peacock, forwarded by Henslow. "My dear Henslow", Peacock's letter read, "Captain Fitz-Roy is going out to survey the southern coast of Tierra del Fuego, and afterwards to visit many of the South Sea Islands, and to return by the Indian Archipelago... An offer has been made to me to recommend a proper person to go out as a naturalist with the expedition. He will be treated with every consideration. The Captain is a young man of very pleasant manners (a nephew of the Duke of Grafton), of great zeal in his profession and highly spoken of..."

In forwarding this letter to Darwin, Henslow added: "I have stated that I consider you to be the best qualified person I know of who is likely to undertake such a situation... The voyage is to last two years and if you take plenty of books with you, anything you please may be done... In short, I suppose that there never was a finer chance for a young man of zeal and spirit..."

Darwin was beside himself with joy at this chance to follow in the footsteps of his hero, Alexander von Humboldt; but his plans were immediately squelched by the opposition of his father, who considered it "a wild scheme", unsuitable for a future clergyman. "If you can find any man of common sense who advises you to go", his father added, "I will give my consent". Crushed by his father's refusal, Charles Darwin visited his uncle's family. Darwin's favorite "Uncle Jos" was the son of the famous potter, Josiah Wedgewood, and the nearby Wedgewood estate at Maer was always a more relaxing place for him than his own home - a relief from the overpowering presence of his father. (His uncle's many attractive daughters may also have had something to do with Darwin's fondness for Maer.)

The Wedgewood family didn't seem to think that sailing on the *Beagle* as naturalist would be a "wild scheme", and Darwin's Uncle Jos offered to drive him over to see whether the verdict could be changed. "My father always maintained that my uncle was one of the most sensible men in the world", Darwin wrote later, "and he at once consented in the kindest manner". Darwin had been rather extravagant while at Cambridge, and to console his father he said: "I should be deuced clever to spend more than my allowance whilst on board the *Beagle*". His father answered with a smile: "But they tell me you are very clever".

Aboard the Beagle

Thus it happened that on December 27, 1831, Charles Darwin sailed from Devonport on *H.M.S. Beagle*, a small brig of the British navy. The *Beagle's* commander, Captain FitzRoy, was twenty-seven years old (four years older than Darwin), but he was already an excellent and experienced sailor. He had orders to survey the South American coast and to carry a chain of chronological measurements around the world. It was to be five years before the Beagle returned to England.

As the brig plowed through rough winter seas, Darwin lay in his hammock, miserably seasick and homesick, trying bravely to read a new book which Henslow had given to him as a sending-off present: Sir Charles Lyell's *Principles of Geology*. It was an exciting and revolutionary book — so revolutionary, in fact, that Henslow had found it necessary to warn Darwin not to believe Lyell's theories, but only to trust his observations. According to Lyell, "No causes have ever acted (in geology) but those which now are acting, and they have never acted with different degrees of energy from that which they now exert"[2].

Lyell's hypothesis was directly opposed to the Catastrophist school of geology, a school which included deeply religious men like Cuvier, Henslow and Sedgwick, as well as most other naturalists of the time. The Catastrophists admitted that geological evidence shows the earth to be much older than the six thousand years calculated on the basis of the Bible, but they explained this by saying that the Bible describes only the most recent era. Before this, according to the Catastrophists, life on earth had been created many times, and just as many times destroyed by cataclysms like Noah's flood.[3] In this way they explained the fossils embedded in ancient rocks: These they believed were the remains of antediluvian creatures destroyed by the wrath of God. The Swiss naturalist Charles Bonnet (1720–1793) even predicted a future catastrophe after which apes would become men and men would become angels. The Catastrophists believed that periodic cataclysms had created the earth's great mountain ranges, deserts and oceans.

Lyell's book contradicted this whole picture. He believed the earth to be immensely old, and asserted that over thousands of millions of years, the

---

[2] This is the famous Principle of Uniformitarianism first formulated by Hutton and later developed in detail by Lyell.

[3] One group of Catastrophists, the Neptunists, believed that gigantic floods shaped the earth's features. A rival group, the Plutonists, attributed most geological features to volcanic action, rather than flood.

same slow changes which we can still see taking place have accumulated to produce the earth's great geological features. Over long ages, Lyell believed, gradual changes in the level of the land built up even the highest mountain ranges, while the slow action of rain and frost cut the peaks into valleys and planes.

By the time the *Beagle* reached the volcanic island of St. Jago, Darwin had become ardently converted to Lyell's "wonderfully superior method of treating geology"; and after studying the structure of the island, he realized that he could understand it on the basis of Lyell's principles. The realization that he might perhaps write a book on the geology of the various countries visited by the *Beagle* made Darwin's spirits soar; and he was thrilled also by the sight of so many totally new species of birds, insects and flowers.

"It has been a glorious day", he wrote, "like giving a blind man eyes: He is overwhelmed by what he sees and cannot easily comprehend it". Later, when the *Beagle* reached Brazil, Darwin was greatly moved by the experience of standing for the first time among the cathedral-like arches of a tropical rain forest. "My mind has been, since leaving England, in a perfect hurricane of delight and astonishment", he wrote, "The glorious pleasure of walking amongst such flowers and such trees cannot be comprehended by those who have not experienced it... Here (the naturalist) suffers a pleasant nuisance of being fairly tied to the spot by some new and wondrous creature... twiners entwining twiners - tresses like hair - beautiful Lepidoptera - silence - hosanna... I am at present fitted for nothing but to read Humboldt: He is like another sun, illuminating all that I behold".

While Captain FitzRoy sailed the *Beagle* slowly southward towards Tierra del Fuego, Darwin followed the ship on horseback, studying the geology of the Argentine Pampas and collecting specimens to send back to Cambridge. Darwin's companions on these expeditions were gauchos, wild Argentine horsemen, expert at throwing the lazo and bolas while galloping at full speed. On one of his rides across the Pampas, Darwin came across the bones of an enormous animal, half buried in a bank of mud and ancient seashells. In a state of great excitement he dug in the surrounding area, and in a few days he succeeded in unearthing the remains of nine huge extinct animals. He was struck by the fact that the bones resembled those of various living South American animals, except for their colossal size. Among them was a guanaco (a wild llama) as big as a camel, a huge armadillo-like creature and a giant sloth-like animal, both as big as elephants. What was the relationship between these extinct animals and living South American species? This problem was to haunt Darwin for many years.

On its way to Tierra del Fuego, the *Beagle* stopped at the Falkland Islands, and Darwin was fascinated by the strange flightless "steamer" ducks found there. He noted that their wings were too small and weak to allow flight. The ducks seemed to paddle with their right and left wings alternately in swimming along the surface of the water; and in this way they were able to move very fast. Darwin reflected that in the South American region there were three species of birds which used their wings for purposes other than flight: the steamer ducks used their wings as paddles, penguins used them as fins, and ostriches used them as sails. Did the ancestors of these birds use their wings for flying? Had the function of the wings changed over a period of time?

On the Falkland Islands, Darwin also noticed that the wild horses had become much smaller than their ancestors, the European horses released there almost three centuries earlier. If the Falkland horses had become noticeably smaller during only a few centuries, then perhaps, over millions of years, the giant armadillo and sloth could have shrunk from the monstrous size of the bones discovered by Darwin to their present size. Perhaps also the wings of the steamer duck, the penguin and the ostrich had become smaller, so that the birds had lost the power of flight. Recalling Lyell's belief in the immense age of the earth, Darwin began to wonder whether small changes, continued over long periods of time, could ultimately produce large changes in living things as well as in geology.

The *Beagle* rounded Cape Horn, lashed by freezing waves so huge that it almost foundered. After the storm, when the brig was anchored safely in the channel of Tierra del Fuego, Darwin noticed how a Fuegian woman stood for hours and watched the ship, while sleet fell and melted on her naked breast, and on the new-born baby she was nursing. He was struck by the remarkable degree to which the Fuegians had adapted to their frigid environment, so that they were able to survive with almost no shelter, and with no clothes except a few stiff, untanned animal skins, which hardly covered them, in weather which would have killed ordinary people.

In 1835, as the *Beagle* made its way slowly northward, Darwin had many chances to explore the Chilean coast — a spectacularly beautiful country, shadowed by towering ranges of the Andes. On January 15, the watch on the *Beagle* noticed something resembling a large star, which gradually increased in size and brilliance. Looking through their telescope, the officers of the *Beagle* could see that the volcano of Osorno was erupting. Darwin was later surprised to learn that on the same night several other volcanos, spread along three thousand miles of coast, had simultaneously erupted.

On February 20, Darwin felt the shock of a severe earthquake, which totally destroyed the towns of Talcahuano and Concepcion. Near the Bay of Concepcion, he could see that the level of the land had been raised three feet by the earthquake; and on the nearby island of St. Maria, Captain FitzRoy found banks of decaying mussel-shells on rocks ten feet above the water line. After the earthquake, it was easy for Darwin to visualize the process by which, over millions of years, the Andes had been raised from the ocean. The sea shells which he found high in the mountains showed that even the highest peaks had once been under the Pacific. Later, high in the Andes, Darwin observed the opposing process - the process by which mountain ranges are torn down. Beside a rushing torrent he stood listening to the rattling noise of stones carried downward by the water. "The sound spoke eloquently to the geologist", he wrote, "The thousands and thousands of stones, which striking against each other made one dull uniform sound, were all hurrying in one direction. It was like thinking on time... As often as I have seen beds of mud, sand and shingles, accumulated to the thickness of many thousands of feet, I have felt inclined to exclaim that causes such as present rivers and present beaches could never have ground down and produced such masses. But on the other hand, while listening to the rattling noise of these torrents and calling to mind that whole races of animals have passed away from the face of the earth, and that during this whole period, night and day, these stones have gone rattling in their course, I have thought to myself, can any mountains, any continent, withstand such a waste?"

After charting the Chilian coast, the *Beagle* sailed westward into the Pacific; and on September 15, 1835, the brig arrived at the Galapagos Archipelago, a group of strange volcanic islands about 500 miles from the mainland. Most of the species of plants, birds and animals which Darwin found on these islands were aboriginal species, found nowhere else in the world; yet in studying them he was continually reminded of species which he had seen on the South American continent. For example, a group of aboriginal finches which Darwin found on the Galapagos Islands were related to South American finches. The Galapagos finches were later shown to belong to thirteen separate species, all closely similar to each other, but differing in their habits and in the structure of their beaks.[4]

The geology of the islands showed that they had been pushed up from the bed of the sea by volcanic action in fairly recent times. Originally each

---

[4] Darwin was not even aware at the time that they were finches. It was on his return to London that an ornithologist friend identified them, noted their close relationship to an Ecuadorian finch, and Darwin came to understand their significance.

island must have been completely bare of plants and animals. How had it been populated? The fact that the Galapagos species resembled those of the South American mainland made it seem probable to Darwin that the islands had become the home of chance wanderers from the continent. Seeds had perhaps drifted onto the shore and germinated, or perhaps they had been brought to the islands in the stomachs of birds. Land birds, like the Galapagos finches, could have been blown there by storms. Perhaps a flock of a single species of finch had arrived, storm-driven, on the black volcanic shores of the islands. Over the centuries, as the finches multiplied, their beaks could have become adapted to the various forms of food available. "The most curious fact", Darwin wrote later, "is the perfect gradation in the size of the beaks in the various species... Seeing this gradation and diversity in one small, intimately related group of birds, one might really fancy that from an original paucity of birds in this archipelago, one species had been taken and modified for different ends.. Here... we seem to be brought somewhat near to that great fact - that mystery of mysteries - the first appearance of new beings on this earth".

The idea of the gradual modification of species could also explain the fact, observed by Darwin, that the fossil animals of South America were more closely related to African and Eurasian animals than were the living South American species. In other words, the fossil animals of South America formed a link between the living South American species and the corresponding animals of Europe, Asia and Africa. The most likely explanation for this was that the animals had crossed to America on a land bridge which had since been lost, and that they had afterwards been modified.

The *Beagle* continued its voyage westward, and Darwin had a chance to study the plants and animals of the Pacific Islands. He noticed that there were no mammals on these islands, except bats and a few mammals brought by sailors. It seemed likely to Darwin that all the species of the Pacific Islands had reached them by crossing large stretches of water after the volcanic islands had risen from the ocean floor; and this accounted for the fact that so many classes were missing. The fact that each group of islands had its own particular species, found nowhere else in the world, seemed to Darwin to be strong evidence that the species had been modified after their arrival. The strange marsupials of the isolated Australian continent also made a deep impression on Darwin.

Work in London and Down

The *Beagle* was now on its way home, and Darwin impatiently counted the days and miles which separated him from his family and friends. To his sisters he wrote: "I feel inclined to write about nothing else but to tell you, over and over again, how I long to be quietly seated among you"; and in a letter to Henslow he exclaimed: "Oh the degree to which I long to be living quietly, without one single novel object near me! No one can imagine it until he has been whirled around the world, during five long years, in a Ten Gun Brig".

Professor Sedgwick had told Darwin's father that he believed that Charles would take his place among the leading scientific men of England. This encouraging news from home reached Darwin on Ascension Island. "After reading this letter", Darwin wrote, "I clambered over the mountains with a bounding step and made the rocks resound under my geological hammer".

On October 2,1836, the *Beagle* docked at Falmouth, and Darwin, "giddy with joy and confusion", took the first available coach to The Mount, his family's home in Shrewsbury. After a joyful reunion with his family, he visited the Wedgwood estate at Maer, where his Uncle Jos and his pretty cousins were equally impatient to see him. To Henslow he wrote: "I am in the clouds, and neither know what to do or where to go... My chief puzzle is about the geological specimens - who will have the charity to help me in describing their mineralogical nature?"

Soon Darwin found a collaborator and close friend in none other than Sir Charles Lyell, the great geologist whose book had so inspired him. One of Lyell's best characteristics was his warmth in encouraging promising young scientists. Darwin's theory of the formation of coral barrier reefs and atolls had supplanted Lyell's own theory, but far from being offended, Lyell welcomed Darwin's ideas with enthusiasm. According to Lyell's earlier theory, coral atolls are circular in shape because they are based on the circular rims of submerged volcanos. However, Darwin showed that any island gradually sinking beneath the surface of a tropical ocean can develop into an atoll. He showed that the reef-building organisms of the coral are poisoned by the stagnant water of the central lagoon, but they flourish on the perimeter, where new water is constantly brought in by the waves. Darwin was able to use the presence of coral atolls to map whole regions of the Pacific which are gradually sinking. He pointed out that in the subsiding regions there are no active volcanos, while in regions where the land is rising, there is much volcanic activity.

The years between 1836 and 1839 were busy ones for Darwin. He found lodgings in London, and he worked there with Lyell on his geological collection. During these years he edited a five-volume book on the zoological discoveries of the voyage; and in 1839 his *Journal of Researches into the Geology and Natural History of Various Countries Visited by the H.M.S. Beagle* was published. Originally Darwin's journal formed part of a multi-volume work edited by Captain FitzRoy, but the publisher, John Murray, recognized the unusual interest of Darwin's contribution, bought up the copyright, and republished the journal. It immediately became a best-seller, making Darwin famous. Under the shortened title, The *Voyage of the Beagle*, Darwin's journal has been reprinted more than a hundred times.

In 1839 Darwin married his pretty cousin, Emma Wedgwood, the youngest daughter of his much-admired Uncle Jos. She was a charming and light-hearted girl who has studied piano under Chopin. Emma and Charles Darwin were to have ten children together (of whom three were knighted for their contributions to science[5]) and thirty years later he wrote of her: "I can declare that in my whole life I have not heard her utter one word which had rather had been left unsaid."

Darwin was beginning to show signs of the ill health which was to remain with him for the rest of his life, and to escape from the social life of the capital, he moved to the small country town of Down, about 16 miles south of London. Darwin's illness was probably due to a chronic infection — perhaps Chagas disease — picked up in South America. For the remainder of his life, his strength was very limited, and his daily routine at Down followed an unvarying pattern which allowed him to work as much as possible within the limits imposed by his illness. The early mornings were devoted to writing (even Sunday mornings) while correspondence and experimental work were done in the afternoons and scientific reading in the evenings.

The Origin of Species

In 1837 Darwin had begun a notebook on Transmutation of Species. During the voyage of the *Beagle* he had been deeply impressed by the great fossil animals which he had discovered, so like existing South American species

---

[5] Among Darwin's grandchildren were Sir Charles Galton Darwin, a pioneer of relativistic quantum theory, and the artist and author, Gwen Raverat. One of his grand-nephews was the composer, Ralph Vaughn Williams.

except for their gigantic size. Also, as the *Beagle* had sailed southward, he had noticed the way in which animals were replaced by closely allied species. On the Galapagos Islands, he had been struck by the South American character of the unique species found there, and by the way in which they differed slightly on each island.

It seemed to Darwin that these facts, as well as many other observations which he had made on the voyage, could only be explained by assuming that species gradually became modified. The subject haunted him, but he was unable to find the exact mechanism by which species changed. Therefore he resolved to follow the Baconian method, which his friend Sir Charles Lyell had used so successfully in geology. He hoped that by the wholesale collection of all facts related in any way to the variation of animals and plants under domestication and in nature, he might be able to throw some light on the subject. He soon saw that in agriculture, the key to success in breeding new varieties was selection; but how could selection be applied to organisms living in a state of nature?

In October 1838, 15 months after beginning his systematic enquiry, Darwin happened to read Malthus' book on population.[6] After his many years as a naturalist, carefully observing animals and plants, Darwin was very familiar with the struggle for existence which goes on everywhere in nature; and it struck him immediately that under the harsh conditions of this struggle, favorable variations would tend to survive while unfavorable ones would perish. The result would be the formation of new species!

Darwin had at last got a theory on which to work, but he was so anxious to avoid prejudice that he did not write it down. He continued to collect facts, and it was not until 1842 that he allowed himself to write a 35-page sketch of his theory. In 1844 he enlarged this sketch to 230 pages, and showed it to his friend Sir Joseph Hooker, the Director of Kew Botanical Gardens. However, Darwin did not publish his 1844 sketch. Probably he foresaw the storm of bitter hostility which his heretical theory was to arouse. In England at that time, Lamarckian ideas from France were regarded as both scientifically unrespectable and politically subversive. The hierarchal English establishment was being attacked by the Chartist movement, and troops had been called out to suppress large scale riots and to ward off revolution. Heretical ideas which might undermine society were regarded as extremely dangerous. Darwin himself was a respected member of the

---

[6] *An Essay on the Principle of Population, or, A View of its Past and Present Effects, with an Inquiry into our Prospects Respecting its Future Removal or Mitigation of the Evils which it Occasions*, 2nd edn, Johnson, London (1803).

establishment, and he was married to a conservative and devout wife, whose feelings he wished to spare. So he kept his work on species private, confiding his ideas only to Hooker and Lyell.

Instead of publishing his views on evolution, Darwin began an enormous technical study of barnacles, which took him eight years to finish. Hooker had told him that no one had the right to write on the question of the origin of species without first having gone through the detailed work of studying a particular species. Also, barnacles were extremely interesting to Darwin: They are in fact more closely related to shrimps and crabs than to molluscs.

Finally, in 1854, Darwin cleared away the last of his barnacles and began to work in earnest on the transmutation of species through natural selection, arranging the mountainous piles of notes on the subject which he had accumulated over the years. By 1858 he had completed the first part of a monumental work on evolution. If he had continued writing on the same scale, he would ultimately have produced a gigantic, unreadable multivolume opus. Fortunately this was prevented: A young naturalist named Alfred Russell Wallace, while ill with a fever in Malaya, also read Malthus on Population; and in a fit of inspiration he arrived at a theory of evolution through natural selection which was identical with Darwin's! Wallace wrote out his ideas in a short paper with the title: *On the Tendency of Varieties to Depart Indefinitely from the Original Type*. He sent this paper to Darwin with the request that if Darwin thought the paper good, he should forward it to Lyell.

Lyell had for years been urging Darwin to publish his own work on natural selection, telling him that if he delayed, someone else would reach the same conclusions. Now Lyell's warning had come true with a vengeance, and Darwin's first impulse was to suppress all his own work in favor of Wallace. In a letter to Lyell, Darwin wrote: "I would far rather burn my whole book than that he or any other man should think that I had behaved in a paltry spirit". Darwin's two good friends, Lyell and Hooker, firmly prevented this however; and through their intervention a fair compromise was reached: Wallace's paper, together with an extract from Darwin's 1844 sketch on natural selection, were read jointly to the Linnean Society (which listened in stunned silence).

At the urging of Lyell and Hooker, Darwin now began an abstract of his enormous unfinished book. This abstract, entitled *On The Origin of Species by Means of Natural Selection, or The Preservation of Favoured Races in the Struggle for Life*, was published in 1859. It ranks with Newton's *Principia* as one of the two greatest scientific books ever written.

Darwin's *Origin of Species* can still be read with enjoyment and fascination by a modern reader. His style is vivid and easy to read, and almost all of his conclusions are still believed to be true. Darwin begins his great book with a history of evolutionary ideas. He starts with a quotation from Aristotle, who was groping towards the idea of natural selection: "Wheresoever, therefore... all the parts of one whole happened like as if they were made for something, these were preserved, having been appropriately constituted by an internal spontaneity; and wheresoever things were not thus constituted, they perished, and still perish". Darwin lists many others who contributed to evolutionary thought, including the Chevalier de Lamarck, Geoffroy Saint-Hilaire, Alfred Russell Wallace, and his own grandfather, Erasmus Darwin.

Next, Darwin reminds us of the way in which mankind has produced useful races of domestic animals and plants by selecting from each generation those individuals which show any slight favorable variation, and by using these as parents for the next generation. A closely similar process occurs in nature, Darwin tells us: Wild animals and plants exhibit slight variations, and in nature there is always a struggle for existence. This struggle follows from the fact that every living creature produces offspring at a rate which would soon entirely fill up the world if no check ever fell on the growth of population. We often have difficulty in seeing the exact nature of these checks, since living organisms are related to each other and to their environment in extremely complex ways, but the checks must always be present.

Accidental variations which increase an organism's chance of survival are more likely to be propagated to subsequent generations than are harmful variations. By this mechanism, which Darwin called "natural selection", changes in plants and animals occur in nature just as they do under the artificial selection exercised by breeders.

If we imagine a volcanic island, pushed up from the ocean floor and completely uninhabited, we can ask what will happen as plants and animals begin to arrive. Suppose, for example, that a single species of bird arrives on the island. The population will first increase until the environment cannot support larger numbers, and it will then remain constant at this level. Over a long period of time, however, variations may accidentally occur in the bird population which allow the variant individuals to make use of new types of food; and thus, through variation, the population may be further increased.

In this way, a single species "radiates" into a number of sub-species which fill every available ecological niche. The new species produced in this way will be similar to the original ancestor species, although they may be greatly modified in features which are related to their new diet and habits. Thus, for example, whales, otters and seals retain the general structure of land-going mammals, although they are greatly modified in features which are related to their aquatic way of life. This is the reason, according to Darwin, why vestigial organs are so useful in the classification of plant and animal species.

The classification of species is seen by Darwin as a genealogical classification. All living organisms are seen, in his theory, as branches of a single family tree. This is a truly remarkable assertion, since the common ancestors of all living things must have been extremely simple and primitive; and it follows that the marvelous structures of the higher animals and plants, whose complexity and elegance utterly surpasses the products of human intelligence, were all produced, over thousands of millions of years, by random variation and natural selection!

Each structure and attribute of a living creature can therefore be seen as having a long history; and a knowledge of the evolutionary history of the organs and attributes of living creatures can contribute much to our understanding of them. For instance, studies of the evolutionary history of the brain and of instincts can contribute greatly to our understanding of psychology, as Darwin pointed out.

Darwin then discusses the complex networks of relationships between living organisms[7]. For example, he discusses the way in which a certain kind of fly prevents horses, cattle and dogs from becoming feral (i.e., thriving as wild animals) in Paraguay. The fly lays its eggs in the navels of these animals when they are born. If the infestations are untreated, fewer of the newborns survive. In other parts of South America, to the north and south of Paraguay, the flies are less numerous, probably because of the presence of parasitic insects. Hence, Darwin concludes, if insect-eating birds were to decrease in Paraguay, the parasitic insects would increase, and this would lessen the number of navel-frequenting flies. Then cattle and horses would become feral, and this would alter the vegetation, which would affect the insects, and so on in ever-increasing circles of complexity.

Another interesting chain of ecological relationships involves clover, bumble-bees, mice, cats and cat-loving people: Red clover is much more common near to towns than elsewhere. Why should this be so? Darwin's

---

[7] Here we can see Darwin as the founder of the modern discipline of ecology.

explanation is that this type of clover can only be pollinated by bumble-bees. The underground nests of bumble-bees are often destroyed by mice; but near to towns mice are kept in check by cats. Hence, Darwin notes, the presence of cats in a district might determine, through the intervention first of mice and then of bees, the frequency of certain flowers in that district.

Among the many striking observations presented by Darwin to support his theory, are facts related to morphology and embryology. For example, Darwin includes a quotation from the naturalist, von Baer, who stated that he had in his possession two embryos preserved in alcohol, which he had forgotten to label. Von Baer was completely unable to tell by looking at them whether they were embryos of lizards, birds or mammals, since all these species are so similar at an early stage of development.

Darwin also quotes the following passage from G.H. Lewis: "The tadpole of the common Salamander has gills, and passes its existence in the water; but the Salamandra atra, which lives high up in the mountains, brings forth its young full-formed. This animal never lives in the water. Yet if we open a gravid female, we find tadpoles inside her with exquisitely feathered gills; and when placed in water, they swim about like the tadpoles of the common Salamander or water-newt. Obviously this aquatic organization has no reference to the future life of the animal, nor has it any adaptation to its embryonic condition; it has solely reference to ancestral adaptations; it repeats a phase in the development of its progenitors".

Darwin points out that, "...As the embryo often shows us more or less plainly the structure of the less modified and ancient progenitor of the group, we can see why ancient and extinct forms so often resemble in their adult state the embryos of existing species".

Darwin sets forth another line of argument in support of evolution based on "serial homologies" — cases where symmetrically repeated parts of an ancient progenitor have been modified for special purposes in their descendants. For example, the bones which fit together to form the brain case in reptiles, birds and mammals can be seen in fossil sequences to be modified vertebrae of an ancient progenitor. After discussing many examples, Darwin exclaims, "How inexplicable are these cases of serial homologies on the ordinary view of creation! Why should the brain be enclosed in a box composed of such numerous and extraordinarily-shaped pieces of bone?... Why should similar bones have been created to form the wing and leg of a bat, used as they are for totally different purposes, namely walking and fly-ing? Why should one crustacean, which has an extremely complex mouth, formed of many parts, consequently have fewer legs; or conversely, those with many legs have simpler mouths? Why should the sepals, petals, stamens and pistils in each flower,though fitted for such distinct purposes, be all constructed on the same pattern?... On the theory of natural selection we can, to a certain extent, answer these questions.... An indefinite repetition of the same part is the common characteristic of all low or little-specialized forms... We have already seen that parts many times repeated are eminently liable to vary... Consequently such parts, being already present in considerable numbers, and being highly variable, would naturally afford materials for adaption to the most different purposes".

No abstract of Darwin's book can do justice to it. One must read it in the original. He brings forward an overwhelming body of evidence to support his theory of evolution through natural selection; and he closes with the following words:

"It is interesting to contemplate a tangled bank, clothed with many plants of many different kinds, with birds singing on the bushes, with various insects flitting about, and with worms crawling through the damp earth, and to reflect that these elaborately constructed forms, so different from each other, and dependent upon each other in so complex a manner, have all been produced by laws acting around us... There is grandeur in this view of life, with its several powers, having been originally breathed by the Creator into a few forms or into one; and that whilst this planet has gone cycling on according to the fixed law of gravity, from so simple a beginning, endless forms most beautiful and wonderful have been and are being evolved".

The Descent of Man

Darwin's *Origin of Species*, published in 1859, was both an immediate success and an immediate scandal. Darwin had sent an advance copy of his book to The Times to be reviewed; and because of the illness of the usual reviewer, T.H. Huxley (1825–1895) was asked to comment on the book. Huxley, who was one of the most brilliant zoologists of the period, immediately recognized the validity and importance of Darwin's work and exclaimed: "How exceedingly stupid not to have thought of that!" He wrote a long and favorable review for The Times, and partly as a result of this review, the first edition of *The Origin of Species* (1200 copies) was sold out on the day of publication. A second edition, published six weeks later, also sold out quickly; and new editions, reprintings and translations have been published ever since in a steady stream.

Darwin had avoided emphasizing the emotionally-charged subject of man's ancestry, but he did not think that it would be honest to conceal his belief that the human race belongs to the same great family which includes all other living organisms on earth. As a compromise, he predicted in a single sentence that through studies of evolution "light would be thrown on the origin of man and his history". This single sentence, and the obvious implications of Darwin's book, were enough to create a storm of furious opposition. One newspaper commented that "society must fall to pieces if Darwinism be true".

The storm of scandalized opposition was still growing in June 1860, when three anti-Darwinian papers were scheduled for reading at an open meeting of the British Association for the Advancement of Science at Oxford. The meeting hall was packed with 700 people as Samuel Wilberforce, Bishop of Oxford, took the floor to "smash Darwin". Darwin himself was too ill (or too diffident) to be present, but T.H. Huxley had been persuaded to attend the meeting to defend Darwin's ideas. After savagely attacking Darwin for half an hour, the bishop turned to Huxley and asked sneeringly, "Is it through your grandfather or your grandmother that you claim to be descended from an ape?"

Huxley, who was 35 at the time and at the height of his powers, rose to answer the bishop. He first gave scientific answers, point by point, to the objections which had been made to the theory of evolution. Finally, regarding the bishop's question about his ancestry, Huxley said: "If I had to choose between a poor ape for an ancestor and a man, highly endowed by nature and of great influence, who used those gifts to introduce ridicule into a scientific discussion and to discredit humble seekers after truth, I would affirm my preference for the ape". Huxley later recalled: "My retort caused inextinguishable laughter among the people".

Pandemonium broke out in the hall. Lady Brewster fainted, and Admiral FitzRoy, the former captain of the *Beagle*, rose to his feet, lifting a Bible in his hand, exclaiming that the Scriptures are the only reliable authority. Had he known Darwin's true nature, FitzRoy said, he would never have allowed him to sail on board the *Beagle*. As *Macmillan's Magazine* reported later, "Looks of bitter hatred were directed to those who were on Darwin's side". However, later that evening, in the discussions of the events of the day which took place in the Oxford colleges, Darwin's ideas were given a surprisingly fair hearing.

The debate at Oxford marked the turning-point in the battle over evolution. After that, Huxley and Hooker defended Darwin's theories with increasing success in England, while in Germany most of the prominent biologists, led by Professor Ernst Haeckel, were soon on Darwin's side. In America the theory of evolution was quickly accepted by almost all of the younger scientists, despite the opposition of the aging "creationist" Louis Agassiz. However, opposition from religious fundamentalists continued in most parts of America, and in Tennessee a school teacher named John T. Scopes was brought to trial for teaching the theory of evolution. He was prosecuted by the orator and three-time presidential candidate William Jennings Bryan, and defended by the brilliant Chicago lawyer Clarence Darrow. In this famous "Monkey Trial", Scopes was let off with a small fine, but the anti-evolution laws remained in force. It was only in 1968 that the State Legislature of Tennessee repealed its laws against the teaching of evolution[8].

In 1863 Huxley, who was not afraid of controversy, published a book entitled *Evidences of Man's Place in Nature*, and this was followed in 1871 by Darwin's book *The Descent of Man*. Huxley and Darwin brought forward a great deal of evidence to show that human beings are probably descended from an early ape-like primate which is now extinct. Darwin believed that the early stages of human evolution took place in Africa[9]. In order to show that men and apes represent closely-related branches of the same family tree, Darwin and Huxley stressed the many points of similarity — resemblances in structure, reproduction, development, psychology and behavior, as well as susceptibility to the same parasites and diseases.

The Expression of Emotions in Man
and Animals — ethology

In The Origin of Species, Charles Darwin devoted a chapter to the evolution of instincts, and he later published a separate book on *The Expression of Emotion in Man and Animals*. Because of these pioneering studies, Darwin is considered to be the founder of the science of ethology — the study of inherited behavior patterns.

Behind Darwin's work in ethology is the observation that instinctive behavior patterns are just as reliably inherited as morphological character-

---

[8] In 1999, the Kansas State School Board removed biological evolution from the curriculum followed by students within the state. Furthermore, cosmology was also removed from the curriculum because it presents evidence that the earth is extremely old, thus supporting evolution. Fortunately, the 1999 decision has now been reversed.

[9] This guess has been confirmed by the recent discoveries of Broom, Dart and the Leakey family, among many others.

istics. Darwin was also impressed by the fact that within a given species, behavior patterns have some degree of uniformity, and the fact that the different species within a family are related by similarities of instinctive behavior, just as they are related by similarities of bodily form. For example, certain elements of cat-like behavior can be found among all members of the cat family; and certain elements of dog-like or wolf-like behavior can be found among all members of the dog family. On the other hand, there are small variations in instinct among the members of a given species. For example, not all domestic dogs behave in the same way.

"Let us look at the familiar case of breeds of dogs", Darwin wrote in *The Origin of Species*, "It cannot be doubted that young pointers will sometimes point and even back other dogs the very first time they are taken out; retrieving is certainly in some degree inherited by retrievers; and a tendency to run round, instead of at, a flock of sheep by shepherd dogs. I cannot see that these actions, performed without experience by the young, and in nearly the same manner by each individual, and without the end being known — for the young pointer can no more know that he points to aid his master than the white butterfly knows why she lays her eggs on the leaf of the cabbage — I cannot see that these actions differ essentially from true instincts..."

"How strongly these domestic instincts habits and dispositions are inherited, and how curiously they become mingled, is well shown when different breeds of dogs are crossed. Thus it is known that a cross with a bulldog has affected for many generations the courage and obstinacy of greyhounds; and a cross with a greyhound has given to a whole family of shepherd dogs a tendency to hunt hares..."

Darwin believed that in nature, desirable variations of instinct are propagated by natural selection, just as in the domestication of animals, favorable variations of instinct are selected and propagated by kennelmen and stock breeders. In this way, according to Darwin, complex and highly developed instincts, such as the comb-making instinct of honey-bees, have evolved by natural selection from simpler instincts, such as the instinct by which bumble bees use their old cocoons to hold honey and sometimes add a short wax tube.

The study of inherited behavior patterns in animals was continued in the 20th century by such researchers as Nikolaas Tinbergen, Konrad Lorenz and Karl von Frisch, three scientists who shared the first Nobel Prize ever awarded in the field of ethology. Among the achievements for which Tinbergen is famous are his classic studies of instinct in herring gulls. He noticed that the newly-hatched chick of a herring gull pecks at the beak of its parent, and this signal causes the parent gull to regurgitate food into the gaping beak of the chick. Tinbergen wondered what signal causes the chick to initiate this response by pecking at the beak of the parent gull. Therefore he constructed a series of models of the parent in which certain features of the adult gull were realistically represented while other features were crudely represented or left out entirely. He found by trial and error that the essential signal to which the chick responds is the red spot on the tip of its parent's beak. Models which lacked the red spot produced almost no response from the young chick, although in other respects they were realistic models; and the red spot on an otherwise crude model would make the chick peck with great regularity.

Tinbergen called this type of signal a "sign stimulus". He found by further studies that he could produce an even more frantic response from the young chick by replacing the red spot by several concentric black circles on a white background, a sign stimulus which he called "super-normal"

In his 1978 book on *Animal Behavior*, Tinbergen pointed out that the features of baby animals, with their large foreheads, round cheeks, and round eyes, all have a characteristic "baby" look. This, Tinbergen wrote, is a sign stimulus which draws a protective response from adults; and he calls attention to the exaggerated "baby" look of some of Walt Disney's animals as an example of a super-normal sign stimulus. Another example of a super-normal sign stimulus, Tinbergen wrote, is the red lipstick and dark eye makeup sometimes used by women.

In the case of a newly-hatched herring gull chick pecking at the red spot on the beak of its parent, the program in the chick's brain must be entirely genetically determined, without any environmental component at all. Learning cannot play a part in this behavioral pattern, since the pattern is present in the young chick from the very moment when it breaks out of the egg. On the other hand (Tinbergen pointed out) many behavioral patterns in animals and in man have both an hereditary component and an environmental component. Learning is often very important, but learning seems to be built on a foundation of genetic predisposition.

To illustrate this point, Tinbergen called attention to the case of sheep-dogs, whose remote ancestors were wolves. These dogs, Tinbergen tells us, can easily be trained to drive a flock of sheep towards the shepherd. However, it is difficult to train them to drive the sheep away from their master. Tinbergen explained this by saying that the sheep-dogs regard the shepherd as their "pack leader"; and since driving the prey towards the pack leader is part of the hunting instinct of wolves, it is easy to teach the dogs this maneuver. However, driving the prey away from the pack leader would not make sense for wolves hunting in a pack; it is not part of the instinctive makeup of wolves, nor is it a natural pattern of behavior for their remote descendants, the sheep-dogs.

Tinbergen also tells us that a Welsh shepherd who wishes to discipline his dog often bites it in the ear; and this is an extremely effective method of enforcing discipline with dogs. To explain the effectiveness of the ear bite, Tinbergen reminds his readers that the leader of a pack of wolves disciplines his subordinates by biting their ears.

As a further example of the fact that learning is usually built on a foundation of genetic predisposition, Tinbergen mentions the ease with which human babies learn languages. The language learned is determined by the baby's environment; but astonishing ease with which a human baby learns to speak and understand implies a large degree of genetic predisposition.

Suggestions for further reading

(1) Sir Julian Huxley and H.B.D. Kettlewell, *Charles Darwin and his World*, Thames and Hudson, London, (1965).

(2) Allan Moorehead, *Darwin and the Beagle*, Penguin Books Ltd., (1971).

(3) Francis Darwin (editor), *The Autobiography of Charles Darwin and Selected Letters*, Dover, New York, (1958).

(4) Charles Darwin, *The Voyage of the Beagle*, J.M. Dent and Sons Ltd., London, (1975).

(5) Charles Darwin, *The Origin of Species*, Collier MacMillan, London, (1974).

(6) Charles Darwin, *The Expression of Emotions in Man and Animals*, The University of Chicago Press (1965).

(7) D.W. Forest, *Francis Galton, The Life and Work of a Victorian Genius*, Paul Elek, London (1974).

(8) Ruth Moore, *Evolution*, Time-Life Books (1962).

(9) L. Barber, *The Heyday of Natural History: 1820-1870*, Doubleday and Co., Garden City, New York, (1980).

(10) A. Desmond, *Huxley*, Addison Wesley, Reading, Mass., (1994).

(11) R. Owen, (P.R. Sloan editor), *The Hunterian Lectures in Comparative Anatomy*, May-June, 1837, University of Chicago Press, (1992).

(12) C. Nichols, *Darwinism and the social sciences*, Phil. Soc. Scient. 4, 255-277 (1974).

(13) M. Ruse, *The Darwinian Revolution*, University of Chicago Press, (1979).

(14) A. Desmond and J. Moore, *Darwin*, Penguin Books, (1992).

(15) R. Dawkins, *The Extended Phenotype*, Oxford University Press, (1982).

(16) R. Dawkins, *The Blind Watchmaker*, W.W. Norton, (1987).

(17) R. Dawkins, *River out of Eden: A Darwinian View of Life*, Harper Collins, (1995).

(18) R. Dawkins, *Climbing Mount Improbable*, W.W. Norton, (1996).

(19) S.J. Gould, *Ever Since Darwin*, W.W. Norton, (1977).

(20) S.J. Gould, *The Panda's Thumb*, W.W. Norton, (1980).

(21) S.J. Gould, *Hen's Teeth and Horse's Toes*, W.W. Norton, (1983).

(22) S.J. Gould, *The Burgess Shale and the Nature of History*, W.W. Norton, (1989).

(23) R.G.B. Reid, *Evolutionary Theory: The Unfinished Synthesis*, Croom Helm, (1985).

(24) M. Ho and P.T. Saunders, editors, *Beyond Neo-Darwinism: An Introduction to a New Evolutionary Paradigm*, Academic Press, London, (1984).

(25) J.Maynard Smith, *Did Darwin Get it Right? Essays on Games, Sex and Evolution*, Chapman and Hall, (1989).

(26) E. Sober, *The Nature of Selection: Evolutionary Theory in Philosophical Focus*, University of Chicago Press, (1984).

(27) B.K. Hall, *Evolutionary Developmental Biology*, Chapman and Hall, London, (1992).

(28) J. Thompson, *Interaction and Coevolution*, Wiley and Sons, (1982).

(29) N. Tinbergen, *The Study of Instinct*, Oxford University Press, (1951).

(30) N. Tinbergen, *Social Behavior in Animals*, Methuen, London, (1953).

(31) N. Tinbergen, *The Animal in its World: Explorations of an Ethologist*, Allan and Unwin, London, (1973).

(32) K. Lorenz, *On the evolution of behavior*, Scientific American, December, (1958).

(33) K. Lorenz, *Studies in Animal and Human Behavior. I and II.*, Harvard University Press, (1970) and (1971).

(34) P.H. Klopfer and J.P. Hailman, *An Introduction to Animal Behavior: Ethology's First Century*, Prentice-Hall, New Jersey, (1969).

(35) J. Jaynes, *The historical origins of "Ethology" and "Comparative Psychology"*, Anim. Berhav. 17, 601-606 (1969).

(36) W.H. Thorpe, *The Origin and Rise of Ethology: The Science of the*

Natural Behavior of Animals, Heinemann, London, (1979).

(37) R.A. Hinde, *Animal Behavior: A Synthesis of Ethological and Comparative Psychology*, McGraw-Hill, New York, (1970).

(38) J.H. Crook, editor, *Social Behavior in Birds and Mammals*, Academic Press, London, (1970).

(39) P. Ekman, editor, *Darwin and Facial Expression*, Academic Press, New York, (1973).

(40) P. Ekman, W.V. Friesen and P. Ekworth, *Emotions in the Human Face*, Pergamon, New York, (1972).

(41) N. Burton Jones, editor, *Ethological Studies of Child Behavior*, Cambridge University Press, (1975).

(42) M. von Cranach, editor, *Methods of Inference from Animals to Human Behavior*, Chicago/Mouton, Haag, (1976); Aldine, Paris, (1976).

(43) K. Lorenz, *On Aggression*, Bantam Books, (1977).

(44) I. Eibl-Eibesfeld, *Ethology, The Biology of Behavior*, Holt, Rinehart and Winston, New York, (1975).

(45) P.P.G. Bateson and R.A. Hinde, editors, *Growing Points in Ethology*, Cambridge University Press, (1976).

(46) J. Bowlby, *By ethology out of psychoanalysis: An experiment in interbreeding, Animal Behavior*, **28** , 649-656 (1980).

(47) B.B. Beck, *Animal Tool Behavior*, Garland STPM Press, New York, (1980).

(48) R. Axelrod, *The Evolution of Cooperation*, Basic Books, New York, (1984).

(49) J.D. Carthy and F.L. Ebling, *The Natural History of Aggression*, Academic Press, New York, (1964)

(50) D.L. Cheney and R.M. Seyfarth, *How Monkeys See the World: Inside the Mind of Another Species*, University of Chicago Press, (1990).

(51) F. De Waal, *Chimpanzee Politics*, Cape, London, (1982).

(52) M. Edmunds, *Defense in Animals*, Longman, London, (1974).

(53) R.D. Estes, *The Behavior Guide to African Mammals*, University of California Press, Los Angeles, (1991).

(54) R.F. Ewer, *Ethology of Mammals*, Logos Press, London, (1968).

Chapter 3

MOLECULAR BIOLOGY AND EVOLUTION

Classical genetics

Charles Darwin postulated that natural selection acts on small inheritable variations in the individual members of a species. His opponents objected that these slight variations would be averaged away by interbreeding. Darwin groped after an answer to this objection, but he did not have one. However, unknown to Darwin, the answer had been uncovered several years earlier by an obscure Augustinian monk, Gregor Mendel, who was born in Silesia in 1822, and who died in Bohemia in 1884.

Mendel loved both botany and mathematics, and he combined these two interests in his hobby of breeding peas in the monastery garden. Mendel carefully self-pollinated his pea plants, and then wrapped the flowers to prevent pollination by insects. He kept records of the characteristics of the plants and their offspring, and he found that dwarf peas always breed true — they invariably produce other dwarf plants. The tall variety of pea plants, pollinated with themselves, did not always breed true, but Mendel succeeded in isolating a strain of true-breeding tall plants which he inbred over many generations.

Next he crossed his true-breeding tall plants with the dwarf variety and produced a generation of hybrids. All of the hybrids produced in this way were tall. Finally Mendel self-pollinated the hybrids and recorded the characteristics of the next generation. Roughly one quarter of the plants in this new generation were true-breeding tall plants, one quarter were true-breeding dwarfs, and one half were tall but not true-breeding.

Gregor Mendel had in fact discovered the existence of dominant and recessive genes. In peas, dwarfism is a recessive characteristic, while tallness is dominant. Each plant has two sets of genes, one from each parent. Whenever the gene for tallness is present, the plant is tall, regardless of whether it also has a gene for dwarfism. When Mendel crossed the pure-breeding dwarf plants with pure-breeding tall ones, the hybrids received one type of gene from each parent. Each hybrid had a tall gene and a dwarf gene; but the tall gene was dominant, and therefore all the hybrids were tall. When the hybrids were self-pollinated or crossed with each other, a genetic lottery took place. In the next generation, through the laws of chance, a quarter of the plants had two dwarf genes, a quarter had two tall genes, and half had one of each kind.

Mendel published his results in the *Transactions of the Brünn Natural History Society* in 1865, and no one noticed his paper[1]. At that time, Austria was being overrun by the Prussians, and people had other things to think about. Mendel was elected Abbot of his monastery; he grew too old and fat to bend over and cultivate his pea plants; his work on heredity was completely forgotten, and he died never knowing that he would one day be considered to be the founder of modern genetics.

In 1900 the Dutch botanist named Hugo de Vries, working on evening primroses, independently rediscovered Mendel's laws. Before publishing, he looked through the literature to see whether anyone else had worked on the subject, and to his amazement he found that Mendel had anticipated his great discovery by 35 years. De Vries could easily have published his own work without mentioning Mendel, but his honesty was such that he gave Mendel full credit and mentioned his own work only as a confirmation of Mendel's laws. Astonishingly, the same story was twice repeated elsewhere in Europe during the same year. In 1900, two other botanists (Correns in Berlin and Tschermak in Vienna) independently rediscovered Mendel's laws, looked through the literature, found Mendel's 1865 paper, and gave him full credit for the discovery.

Besides rediscovering the Mendelian laws for the inheritance of dominant and recessive characteristics, de Vries made another very important discovery: He discovered genetic mutations — sudden unexplained changes of form which can be inherited by subsequent generations. In growing evening primroses, de Vries found that sometimes, but very rarely, a completely new variety would suddenly appear, and he found that the variation could be propagated to the following generations. Actually, mutations had been observed before the time of de Vries. For example, a short-legged mutant sheep had suddenly appeared during the 18th century; and stock-breeders had taken advantage of this mutation to breed sheep that could

---

[1] Mendel sent a copy of his paper to Darwin; but Darwin, whose German was weak, seems not to have read it.

not jump over walls. However, de Vries was the first scientist to study and describe mutations. He noticed that most mutations are harmful, but that a very few are beneficial, and those few tend in nature to be propagated to future generations.

After the rediscovery of Mendel's work by de Vries, many scientists began to suspect that chromosomes might be the carriers of genetic information. The word "chromosome" had been invented by the German physiologist, Walther Flemming, to describe the long, threadlike bodies which could be seen when cells were stained and examined through the microscope during the process of division. It had been found that when an ordinary cell divides, the chromosomes also divide, so that each daughter cell has a full set of chromosomes.

The Belgian cytologist, Edouard van Benedin, had shown that in the formation of sperm and egg cells, the sperm and egg receive only half of the full number of chromosomes. It had been found that when the sperm of the father combines with the egg of the mother in sexual reproduction, the fertilized egg again has a full set of chromosomes, half coming from the mother and half from the father. This was so consistent with the genetic lottery studied by Mendel, de Vries and others, that it seemed almost certain that chromosomes were the carriers of genetic information.

The number of chromosomes was observed to be small (for example, each normal cell of a human has 46 chromosomes); and this made it obvious that each chromosome must contain thousands of genes. It seemed likely that all of the genes on a particular chromosome would stay together as they passed through the genetic lottery; and therefore certain characteristics should always be inherited together.

This problem had been taken up by Thomas Hunt Morgan, a professor of experimental zoology working at Colombia University. He found it convenient to work with fruit flies, since they breed with lightning-like speed and since they have only four pairs of chromosomes.

Morgan found that he could raise enormous numbers of these tiny insects with almost no effort by keeping them in gauze-covered glass milk bottles, in the bottom of which he placed mashed bananas. In 1910, Morgan found a mutant white-eyed male fly in one of his milk-bottle incubators. He bred this fly with a normal red-eyed female, and produced hundreds of red-eyed hybrids. When he crossed the red-eyed hybrids with each other, half of the next generation were red-eyed females, a quarter were red-eyed males, and a quarter were white-eyed males. There was not one single white-eyed female! This indicated that the mutant gene for white eyes was on the same chromosome as the gene for the male sex.

As Morgan continued his studies of genetic linkages, however, it became clear that the linkages were not absolute. There was a tendency for all the genes on the same chromosome to be inherited together; but on rare occasions there were "crosses", where apparently a pair of chromosomes broke at some point and exchanged segments. By studying these crosses statistically, Morgan and his "fly squad" were able to find the relative positions of genes on the chromosomes. They reasoned that the probability for a cross to separate two genes should be proportional to the distance between the two genes on the chromosome. In this way, after 17 years of work and millions of fruit flies, Thomas Hunt Morgan and his coworkers were able to make maps of the fruit fly chromosomes showing the positions of the genes.

This work had been taken a step further by Hermann J. Muller, a member of Morgan's "fly squad", who exposed hundreds of fruit flies to X-rays. The result was a spectacular outbreak of man-made mutations in the next generation.

"They were a motley throng", recalled Muller. Some of the mutant flies had almost no wings, others bulging eyes, and still others brown, yellow or purple eyes; some had no bristles, and others curly bristles. Muller's experiments indicated that mutations can be produced by radiation-induced physical damage; and he guessed that such damage alters the chemical structure of genes.

In spite of the brilliant work by Morgan and his collaborators, no one had any idea of what a gene really was.

The structure of DNA

Until 1944, most scientists had guessed that the genetic message was carried by the proteins of the chromosome. In 1944, however, O.T. Avery and his co-workers at the laboratory of the Rockefeller Institute in New York performed a critical experiment, which proved that the material which carries genetic information is not protein, but deoxyribonucleic acid (DNA) — a giant chainlike molecule which had been isolated from cell nuclei by the Swiss chemist, Friedrich Miescher.

Avery had been studying two different strains of pneumococci, the bacteria which cause pneumonia. One of these strains, the S-type, had a smooth coat, while the other strain, the R-type, lacked an enzyme needed for the manufacture of a smooth carbohydrate coat. Hence, R-type pneumococci had a rough appearance under the microscope. Avery and his co-workers were able to show that an extract from heat-killed S-type pneumococci could convert the living R-type species permanently into S-type; and they also showed that this extract consisted of pure DNA.

In 1947, the Austrian-American biochemist, Erwin Chargaff, began to study the long, chainlike DNA molecules. It had already been shown by Levine and Todd that chains of DNA are built up of four bases: adenine (A), thymine (T), guanine (G) and cytosine (C), held together by a sugar-phosphate backbone. Chargaff discovered that in DNA from the nuclei of living cells, the amount of A always equals the amount of T; and the amount of G always equals the amount of C.

When Chargaff made this discovery, neither he nor anyone else understood its meaning. However, in 1953, the mystery was completely solved by Rosalind Franklin and Maurice Wilkins at Kings College, London, together with James Watson and Francis Crick at Cambridge University. By means of X-ray diffraction techniques, Wilkins and Franklin obtained crystallographic information about the structure of DNA. Using this information, together with Linus Pauling's model-building methods, Crick and Watson proposed a detailed structure for the giant DNA molecule.

The discovery of the molecular structure of DNA was an event of enormous importance for genetics, and for biology in general. The structure was a revelation! The giant, helical DNA molecule was like a twisted ladder: Two long, twisted sugar-phosphate backbones formed the outside of the ladder, while the rungs were formed by the base pairs, A, T, G and C. The base adenine (A) could only be paired with thymine (T), while guanine (G) fit only with cytosine (C). Each base pair was weakly joined in the center by hydrogen bonds — in other words, there was a weak point in the center of each rung of the ladder — but the bases were strongly attached to the sugar-phosphate backbone. In their 1953 paper, Crick and Watson wrote:

"It has not escaped our notice that the specific pairing we have postulated suggests a possible copying mechanism for genetic material". Indeed, a sudden blaze of understanding illuminated the inner workings of heredity, and of life itself.

If the weak hydrogen bonds in the center of each rung were broken, the ladderlike DNA macromolecule could split down the center and divide into two single strands. Each single strand would then become a template for the formation of a new double-stranded molecule.

Because of the specific pairing of the bases in the Watson-Crick model of DNA, the two strands had to be complementary. T had to be paired with A, and G with C. Therefore, if the sequence of bases on one strand was (for example) TTTGCTAAAGGTGAACCA... , then the other strand necessarily had to have the sequence AAACGATTTCCACTTGGT... The Watson-Crick model of DNA made it seem certain that all the genetic information needed for producing a new individual is coded into the long, thin, double-stranded DNA molecule of the cell nucleus, written in a four-letter language whose letters are the bases, adenine, thymine, guanine and cytosine.

The solution of the DNA structure in 1953 initiated a new kind of biology — molecular biology. This new discipline made use of recently-discovered physical techniques — X-ray diffraction, electron microscopy, electrophoresis, chromatography, ultracentrifugation, radioactive tracer techniques, autoradiography, electron spin resonance, nuclear magnetic resonance and ultraviolet spectroscopy. In the 1960's and 1970's, molecular biology became the most exciting and rapidly-growing branch of science.

Protein structure

In England, J.D. Bernal and Dorothy Crowfoot Hodgkin pioneered the application of X-ray diffraction methods to the study of complex biological molecules. In 1949, Hodgkin determined the structure of penicillin; and in 1955, she followed this with the structure of vitamin B12. In 1960, Max Perutz and John C. Kendrew obtained the structures of the blood proteins myoglobin and hemoglobin. This was an impressive achievement for the Cambridge crystallographers, since the hemoglobin molecule contains roughly 12,000 atoms.

The structure obtained by Perutz and Kendrew showed that hemoglobin is a long chain of amino acids, folded into a globular shape, like a small, crumpled ball of yarn. They found that the amino acids with an affinity for water were on the outside of the globular molecule; while the amino acids for which contact with water was energetically unfavorable were hidden on the inside. Perutz and Kendrew deduced that the conformation of the protein — the way in which the chain of amino acids folded into a 3-dimensional structure — was determined by the sequence of amino acids in the chain.

In 1966, D.C. Phillips and his co-workers at the Royal Institution in London found the crystallographic structure of the enzyme lysozyme (an egg-white protein which breaks down the cell walls of certain bacteria). Again, the structure showed a long chain of amino acids, folded into a roughly globular shape. The amino acids with hydrophilic groups were on the outside, in contact with water, while those with hydrophobic groups were on the inside. The structure of lysozyme exhibited clearly an active site, where sugar molecules of bacterial cell walls were drawn into a mouth-like opening and stressed by electrostatic forces, so that bonds between the sugars could easily be broken.

Meanwhile, at Cambridge University, Frederick Sanger developed methods for finding the exact sequence of amino acids in a protein chain. In 1945, he discovered a compound (2,4-dinitrofluorobenzene) which attaches itself preferentially to one end of a chain of amino acids. Sanger then broke down the chain into individual amino acids, and determined which of them was connected to his reagent. By applying this procedure many times to fragments of larger chains, Sanger was able to deduce the sequence of amino acids in complex proteins. In 1953, he published the sequence of insulin. This led, in 1964, to the synthesis of insulin.

The biological role and structure of proteins which began to emerge was as follows: A mammalian cell produces roughly 10,000 different proteins. All enzymes are proteins; and the majority of proteins are enzymes — that is, they catalyze reactions involving other biological molecules. All proteins are built from chainlike polymers, whose monomeric sub-units are the following twenty amino acids: glycine, aniline, valine, isoleucine, leucine, serine, threonine, proline, aspartic acid, glutamic acid, lysine, arginine, asparagine, glutamine, cysteine, methionine, tryptophan, phenylalanine, tyrosine and histidine. These individual amino acid monomers may be connected together into a polymer (called a polypeptide) in any order — hence the great number of possibilities. In such a polypeptide, the backbone is a chain of carbon and nitrogen atoms showing the pattern ...-C-C-N-C-C-N-C-C-N-...and so on. The -C-C-N- repeating unit is common to all amino acids. Their individuality is derived from differences in the side groups which are attached to the universal -C-C-N- group.

Some proteins, like hemoglobin, contain metal atoms, which may be oxidized or reduced as the protein performs its biological function. Other proteins, like lysozyme, contain no metal atoms, but instead owe their biological activity to an active site on the surface of the protein molecule. In 1909, the English physician, Archibald Garrod, had proposed a one-gene-one-protein hypothesis. He believed that hereditary diseases are due to the absence of specific enzymes. According to Garrod's hypothesis, damage suffered by a gene results in the faulty synthesis of the corresponding enzyme, and loss of the enzyme ultimately results in the symptoms of the hereditary disease.

In the 1940's, Garrod's hypothesis was confirmed by experiments on the mold, Neurospora, performed at Stanford University by George Beadle and Edward Tatum. They demonstrated that mutant strains of the mold would grow normally, provided that specific extra nutrients were added to their diets. The need for these dietary supplements could in every case be traced to the lack of a specific enzyme in the mutant strains. Linus Pauling later extended these ideas to human genetics by showing that the hereditary disease, sickle-cell anemia, is due to a defect in the biosynthesis of hemoglobin.

RNA and ribosomes

Since DNA was known to carry the genetic message, coded into the sequence of the four nucleotide bases, A, T, G and C, and since proteins were known to be composed of specific sequences of the twenty amino acids, it was logical to suppose that the amino acid sequence in a protein was determined by the base sequence of DNA. The information somehow had to be read from the DNA and used in the biosynthesis of the protein.

It was known that, in addition to DNA, cells also contain a similar, but not quite identical, polynucleotide called ribonucleic acid (RNA). The sugar-phosphate backbone of RNA was known to differ slightly from that of DNA; and in RNA, the nucleotide thymine (T) was replaced by a chemically similar nucleotide, uracil (U). Furthermore, while DNA was found only in cell nuclei, RNA was found both in cell nuclei and in the cytoplasm of cells, where protein synthesis takes place. Evidence accumulated indicating that genetic information is first transcribed from DNA to RNA, and afterwards translated from RNA into the amino acid sequence of proteins.

At first, it was thought that RNA might act as a direct template, to which successive amino acids were attached. However, the appropriate chemical complementarity could not be found; and therefore, in 1955, Francis Crick proposed that amino acids are first bound to an adaptor molecule, which is afterward bound to RNA.

In 1956, George Emil Palade of the Rockefeller Institute used electron microscopy to study subcellular particles rich in RNA (ribosomes). Ribosomes were found to consist of two subunits — a smaller subunit, with a molecular weight one million times the weight of a hydrogen atom, and a larger subunit with twice this weight.

It was shown by means of radioactive tracers that a newly synthesized protein molecule is attached temporarily to a ribosome, but neither of the two subunits of the ribosome seemed to act as a template for protein synthesis. Instead, Palade and his coworkers found that genetic information is carried from DNA to the ribosome by a messenger RNA molecule (mRNA). Electron microscopy revealed that mRNA passes through the ribosome like a punched computer tape passing through a tape-reader. It was found that the adapter molecules, whose existence Crick had postulated, were smaller molecules of RNA; and these were given the name "transfer RNA" (tRNA). It was shown that, as an mRNA molecule passes through a ribosome, amino acids attached to complementary tRNA adaptor molecules are added to the growing protein chain.

The relationship between DNA, RNA, the proteins and the smaller molecules of a cell was thus seen to be hierarchical: The cell's DNA controlled its proteins (through the agency of RNA); and the proteins controlled the synthesis and metabolism of the smaller molecules.

The genetic code

In 1955, Severo Ochoa, at New York University, isolated a bacterial enzyme (RNA polymerase) which was able join the nucleotides A, G, U and C so that they became an RNA strand. One year later, this feat was repeated for DNA by Arthur Kornberg.

With the help of Ochoa's enzyme, it was possible to make synthetic RNA molecules containing only a single nucleotide — for example, one could join uracil molecules into the ribonucleic acid chain, ...U-U-U-U-U-U-... In 1961, Marshall Nirenberg and Heinrich Matthaei used synthetic poly-U as messenger RNA in protein synthesis; and they found that only polyphenylalanine was synthesized. In the same year, Sydney Brenner and Francis Crick reported a series of experiments on mutant strains of the bacteriophage, T4. The experiments of Brenner and Crick showed that whenever a mutation added or deleted either one or two base pairs, the proteins produced by the mutants were highly abnormal and non-functional. However, when the mutation added or subtracted three base pairs, the proteins often were functional. Brenner and Crick concluded that the genetic language has three-letter words (codons). With four different "letters", A, T, G and C, this gives sixty-four possible codons — more than enough to specify the twenty different amino acids.

In the light of the phage experiments of Brenner and Crick, Nirenberg and Matthaei concluded that the genetic code for phenylalanine is UUU

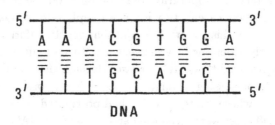

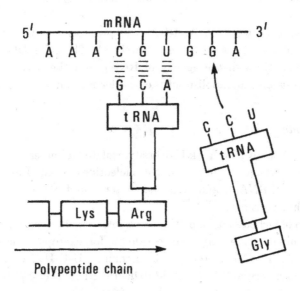

Fig. 3.1   Information coded on DNA molecules in the cell nucleus is transcribed to mRNA molecules. The messenger RNA molecules in turn provide information for the amino acid sequence in protein synthesis.

in RNA and TTT in DNA. The remaining words in the genetic code were worked out by H. Gobind Khorana of the University of Wisconsin, who used other mRNA sequences (such as GUGUGU..., AAGAAGAAG... and GUUGUUGUU...) in protein synthesis. By 1966, the complete genetic code, specifying amino acids in terms of three-base sequences, was known. The code was found to be the same for all species studied, no matter how widely separated they were in form; and this showed that all life on earth belongs to the same family, as postulated by Darwin.

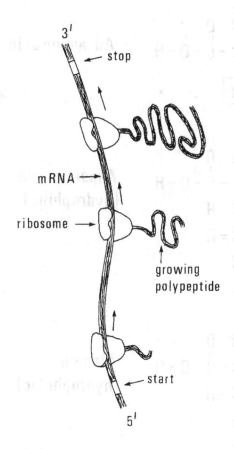

Fig. 3.2   mRNA passes through the ribosome like a punched computer tape passing through a tape-reader.

Genetic engineering

In 1970, Hamilton Smith of Johns Hopkins University observed that when the bacterium Haemophilus influenzae is attacked by a bacteriophage (a virus parasitic on bacteria), it can defend itself by breaking down the DNA of the phage. Following up this observation, he introduced DNA from the bacterium E. coli into H. influenzae. Again the foreign DNA was broken down.

Smith had, in fact, discovered the first of a class of bacterial enzymes which came to be called "restriction enzymes" or "restriction nucleases".

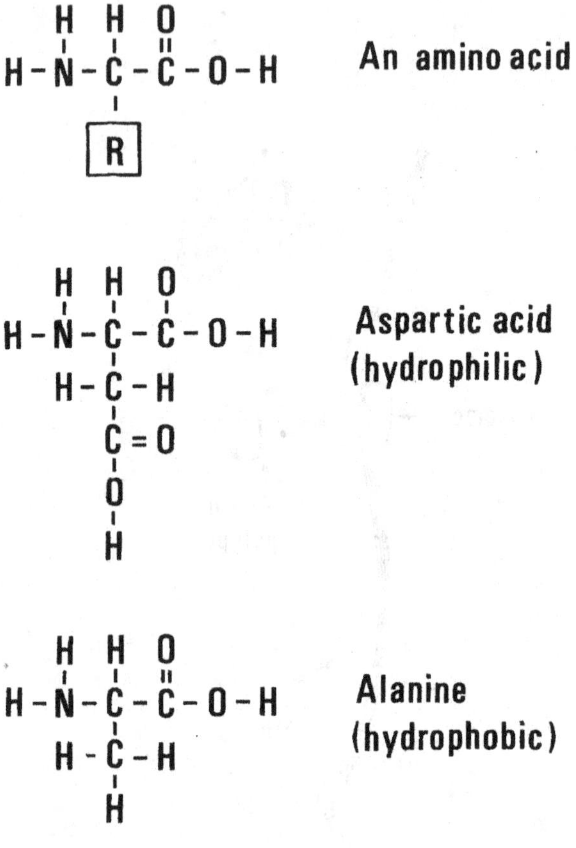

$$
\begin{array}{ccc}
& H & H & O \\
& | & | & \| \\
H - N - & C - & C - O - H \\
& | \\
& \boxed{R}
\end{array}
\qquad \text{An amino acid}
$$

$$
\begin{array}{ccc}
& H & H & O \\
& | & | & | \\
H - N - & C - & C - O - H \\
& H - C - H \\
& C = O \\
& O \\
& H
\end{array}
\qquad
\begin{array}{l}
\text{Aspartic acid} \\
\text{(hydrophilic)}
\end{array}
$$

$$
\begin{array}{ccc}
& H & H & O \\
& | & | & \| \\
H - N - & C - & C - O - H \\
& H - C - H \\
& H
\end{array}
\qquad
\begin{array}{l}
\text{Alanine} \\
\text{(hydrophobic)}
\end{array}
$$

Fig. 3.3   This figure shows aspartic acid, whose residue (R) is hydrophilic, contrasted with alanine, whose residue is hydrophobic.

Almost a hundred other restriction enzymes were subsequently discovered, and each was found to cut DNA at a specific base sequence. Smith's colleague, Daniel Nathans, used the restriction enzymes Hin dll and Hin dill to produce the first "restriction map" of the DNA in a virus.

In 1971 and 1972, Paul Berg, and his co-workers Peter Lobban, Dale Kaiser and David Jackson at Stanford University, developed methods for adding cohesive ends to DNA fragments. Berg and his group used the calf thymus enzyme, terminal transferase, to add short, single-stranded polynucleotide segments to DNA fragments. For example, if they added the single-stranded segment AAAA to one fragment, and TTTT to another,

MOLECULAR BIOLOGY AND EVOLUTION

Table 3.1: The genetic code

TTT=Phe	TCT=Ser	TAT=Tyr	TGT=Cys
TTC=Phe	TCC=Ser	TAC=Tyr	TGC=Cys
TTA=Leu	TCA=Ser	TAA=Ter	TGA=Ter
TTG=Leu	TGC=Ser	TAG=Ter	TGG=Trp
CTT=Leu	CCT=Pro	CAT=His	CGT=Arg
CTC=Leu	CCC=Pro	CAC=His	CGC=Arg
CTA=Leu	CCA=Pro	CAA=Gln	CGA=Arg
CTG=Leu	CGC=Pro	CAG=Gln	CGG=Arg
ATT=Ile	ACT=Thr	AAT=Asn	AGT=Ser
ATC=Ile	ACC=Thr	AAC=Asn	AGC=Ser
ATA=Ile	ACA=Thr	AAA=Lys	AGA=Arg
ATG=Met	AGC=Thr	AAG=Lys	AGG=Arg
GTT=Val	GCT=Ala	GAT=Asp	GGT=Gly
GTC=Val	GCC=Ala	GAC=Asp	GGC=Gly
GTA=Val	GCA=Ala	GAA=Glu	GGA=Gly
GTG=Val	GGC=Ala	GAG=Glu	GGG=Gly

then the two ends joined spontaneously when the fragments were incubated together. In this way, Paul Berg and his group made the first recombinant DNA molecules.

The restriction enzyme Eco RI, isolated from the bacterium E. coli, was found to recognize the pattern, GAATTC, in one strand of a DNA molecule, and the complementary pattern, CTTAAG, in the other strand. Instead of cutting both strands in the middle of the six-base sequence, Eco RI was observed to cut both strands between G and A. Thus, each side of the cut was left with a "sticky end" — a five-base single-stranded segment, attached to the remainder of the double-stranded DNA molecule.

In 1972, Janet Mertz and Ron Davis, working at Stanford University, demonstrated that DNA strands cut with Eco RI could be rejoined by means of another enzyme — a DNA ligase. More importantly, when DNA strands from two different sources were cut with Eco RI, the sticky end of one fragment could form a spontaneous temporary bond with the sticky end of the other fragment. The bond could be made permanent by the addition of DNA ligase, even when the fragments came from different sources. Thus, DNA fragments from different organisms could be joined together.

Bacteria belong to a class of organisms (prokaryotes) whose cells do not have a nucleus. Instead, the DNA of the bacterial chromosome is arranged in a large loop. In the early 1950's, Joshua Lederberg had discovered that bacteria can exchange genetic information. He found that a frequently-exchanged gene, the F-factor (which conferred fertility), was not linked to other bacterial genes; and he deduced that the DNA of the F-factor was not physically a part of the main bacterial chromosome. In 1952, Lederberg coined the word "plasmid" to denote any extrachromosomal genetic system. In 1959, it was discovered in Japan that genes for resistance to antibiotics ·can be exchanged between bacteria; and the name "R-factors" was given to these genes. Like the F-factors, the R-factors did not seem to be part of the main loop of bacterial DNA.

Because of the medical implications of this discovery, much attention was focused on the R-factors. It was found that they are plasmids, small loops of DNA existing inside the bacterial cell but not attached to the bacterial chromosome. Further study showed that, in general, between one percent and three percent of bacterial genetic information is carried by plasmids, which can be exchanged freely even between different species of bacteria.

In the words of the microbiologist, Richard Novick, "Appreciation of the role of plasmids has produced a rather dramatic shift in biologists' thinking about genetics. The traditional view was that the genetic makeup of a species was about the same from one cell to another, and was constant over long periods of time. Now a significant proportion of genetic traits are known to be variable (present in some individual cells or strains, absent in others), labile (subject to frequent loss or gain) and mobile — all because those traits are associated with plasmids or other atypical genetic systems".

In 1973, Herbert Boyer, Stanley Cohen and their co-workers at Stanford University and the University of California carried out experiments in which they inserted foreign DNA segments, cut with Eco RI, into plasmids (also cut with Eco RI). They then resealed the plasmid loops with DNA ligase. Finally, bacteria were infected with the gene-spliced plasmids. The result was a new strain of bacteria, capable of producing an additional protein coded by the foreign DNA segment which had been spliced into the plasmids.

Cohen and Boyer used plasmids containing a gene for resistance to an antibiotic, so that a few gene-spliced bacteria could be selected from a large population by treating the culture with the antibiotic. The selected bacteria, containing both the antibiotic-resistance marker and the foreign DNA, could then be cloned on a large scale; and in this way a foreign gene could be "cloned". The gene-spliced bacteria were chimeras, containing genes from two different species.

The new recombinant DNA techniques of Berg, Cohen and Boyer had revolutionary implications: It became possible to produce many copies of a given DNA segment, so that its base sequence could be determined. With the help of direct DNA-sequencing methods developed by Frederick Sanger and Walter Gilbert, the new cloning techniques could be used for mapping and sequencing genes.

Since new bacterial strains could be created, containing genes from other species, it became possible to produce any protein by cloning the corresponding gene. Proteins of medical importance could be produced on a large scale. Thus, the way was open for the production of human insulin, interferon, serum albumin, clotting factors, vaccines, and protein hormones such as ACTH, human growth factor and leuteinizing hormone.

It also became possible to produce enzymes of industrial and agricultural importance by cloning gene-spliced bacteria. Since enzymes catalyze reactions involving smaller molecules, the production of these substrate molecules through gene-splicing also became possible.

It was soon discovered that the possibility of producing new, transgenic organisms was not limited to bacteria. Gene-splicing was also carried out on higher plants and animals as well as on fungi. It was found that the bacterium Agrobacterium tumefaciens contains a tumor-inducing (Ti) plasmid capable of entering plant cells and producing a crown gall. Genes spliced into the Ti plasmid quite frequently became incorporated in the plant chromosome, and afterwards were inherited in a stable, Mendelian fashion.

Transgenic animals were produced by introducing foreign DNA into embryo-derived stem cells (ES cells). The gene-spliced ES cells were then selected, cultured and introduced into a blastocyst, which afterwards was implanted in a foster-mother. The resulting chimeric animals were bred, and stable transgenic lines selected.

Thus, for the first time, humans had achieved direct control over the process of evolution. Selective breeding to produce new plant and animal varieties was not new — it is one of the oldest techniques of civilization. However, the degree, precision, and speed of intervention which recombinant DNA made possible was entirely new. In the 1970's it became possible to mix the genetic repertoires of different species: The genes of mice and men could be spliced together into new, man-made forms of life!

The Polymerase Chain Reaction

One day in the early 1980's, an American molecular biologist, Kary Mullis, was driving to his mountain cabin with his girl friend. The journey was a long one, and to pass the time, Kary Mullis turned over and over in his mind a problem which had been bothering him: He worked for a California biotechnology firm, and like many other molecular biologists he had been struggling to analyze very small quantities of DNA. Mullis realized that it would be desirable have a highly sensitive way of replicating a given DNA segment — a method much more sensitive than cloning. As he drove through the California mountains, he considered many ways of doing this, rejecting one method after the other as impracticable. Finally a solution came to him; and it seemed so simple that he could hardly believe that he was the first to think of it. He was so excited that he immediately pulled over to the side of the road and woke his sleeping girlfriend to tell her about his idea. Although his girlfriend was not entirely enthusiastic about being wakened from a comfortable sleep to be presented with a lecture on biochemistry, Kary Mullis had in fact invented a technique which was destined to revolutionize DNA technology: the Polymerase Chain Reaction (PCR)[2].

The technique was as follows: Begin with a small sample of the genomic DNA to be analyzed. (The sample may be extremely small — only a few molecules.) Heat the sample to 95 °C to separate the double-stranded DNA molecule into single strands. Suppose that on the long DNA molecule there is a target segment which one wishes to amplify. If the target segment begins with a known sequence of bases on one strand, and ends with a known sequence on the complementary strand, then synthetic "primer" oligonucleotides[3] with these known beginning ending sequences are added in excess. The temperature is then lowered to 50-60 °C, and at the lowered temperature, the "start" primer attaches itself to one DNA strand at the beginning of the target segment, while the "stop" primer becomes attached to the complementary strand at the other end of the target segment. Polymerase (an enzyme which aids the formation of double-stranded DNA) is then added, together with a supply of nucleotides. On each of the original pieces of single-stranded DNA, a new complementary strand is generated with the help of the polymerase. Then the temperature is again raised to

---

[2] The flash of insight didn't take long, but at least six months of hard work were needed before Mullis and his colleagues could convert the idea to reality.

[3] Short segments of single-stranded DNA.

95 °C, so that the double-stranded DNA separates into single strands, and the cycle is repeated.

In the early versions of the PCR technique, the polymerase was destroyed by the high temperature, and new polymerase had to be added for each cycle. However, it was discovered that polymerase from the bacterium Thermus aquaticus would withstand the high temperature. (Thermus aquaticus lives in hot springs.) This discovery greatly simplified the PCR technique. The temperature could merely be cycled between the high and low temperatures, and with each cycle, the population of the target segment doubled, concentrations of primers, deoxynucleotides and polymerase being continuously present.

After a few cycles of the PCR reaction, copies of copies begin to predominate over copies of the original genomic DNA. These copies of copies have a standard length, always beginning on one strand with the start primer, and ending on that strand with the complement of the stop primer.

Two main variants of the PCR technique are possible, depending on the length of the oligonucleotide primers: If, for example, trinucleotides are used as start and stop primers, they can be expected to match the genomic DNA at many points. In that case, after a number of PCR cycles, populations of many different segments will develop. Within each population, however, the length of the replicated segment will be standardized because of the predominance of copies of copies. When the resulting solution is placed on a damp piece of paper or a gel and subjected to the effects of an electric current (electrophoresis), the populations of different molecular weights become separated, each population appearing as a band. The bands are profiles of the original genomic DNA; and this variant of the PCR technique can be used in evolutionary studies to determine the degree of similarity of the genomic DNA of two species.

On the other hand, if the oligonucleotide primers contain as many as 20 nucleotides, they will be highly specific and will bind only to a particular target sequence of the genomic DNA. The result of the PCR reaction will then be a single population, containing only the chosen target segment. The PCR reaction can be thought of as autocatalytic, and as we shall see in the next section, autocatalitic systems play an important role in modern theories of the origin of life.

Theories of chemical evolution towards the origin of life

The possibility of an era of chemical evolution prior to the origin of life entered the thoughts of Charles Darwin, but he considered the idea to be much too speculative to be included in his published papers and books. However, in February 1871, he wrote a letter to his close friend Sir Joseph Hooker containing the following words:

"It is often said that all the conditions for the first production of a living organism are now present, which could ever have been present. But if (and oh what a big if) we could conceive in some warm little pond with all sorts of ammonia and phosphoric salts, light, heat, electricity etc. present, that a protein compound was chemically formed, ready to undergo still more complex changes, at the present day such matter would be instantly devoured, or absorbed, which would not have been the case before living creatures were formed".

The last letter which Darwin is known to have dictated and signed before his death in 1882 also shows that he was thinking about this problem: "You have expressed quite correctly my views", Darwin wrote, "where you said that I had intentionally left the question of the Origin of Life uncanvassed as being altogether ultra vires in the present state of our knowledge, and that I dealt only with the manner of succession. I have met with no evidence that seems in the least trustworthy, in favor of so-called Spontaneous Generation. (However) I believe that I have somewhere said (but cannot find the passage) that the principle of continuity renders it probable that the principle of life will hereafter be shown to be a part, or consequence, of some general law..."

Modern researchers, picking up the problem where Darwin left it, have begun to throw a little light on the problem of chemical evolution towards the origin of life. In the 1930's, J.B.S. Haldane in England and A.I. Oparin in Russia put forward theories of an era of chemical evolution prior to the appearance of living organisms.

In 1924, Oparin published a pamphlet on the origin of life. An expanded version of this pamphlet was translated into English and appeared in 1936 as a book entitled *The Origin of Life on Earth*. In this book, Oparin pointed out that at the time when life originated, conditions on earth were probably considerably different than they are at present: The atmosphere probably contained very little free oxygen, since free oxygen is produced by photosynthesis which did not yet exist. On the other hand, he argued, there were probably large amounts of methane and ammonia in the earth's primitive atmosphere[4]. Thus, before the origin of life, the earth probably had a reducing atmosphere rather than an oxidizing one. Oparin believed that energy-rich molecules could have been formed very slowly by the action of light from the sun. On the present-day earth, bacteria quickly consume energy-rich molecules, but before the origin of life, such molecules could have accumulated, since there were no living organisms to consume them. (This observation is similar to the remark made by Darwin in his 1871 letter to Hooker.)

The first experimental work in this field took place in 1950 in the laboratory of Melvin Calvin at the University of California, Berkeley. Calvin and his co-workers wished to determine experimentally whether the primitive atmosphere of the earth could have been converted into some of the molecules which are the building-blocks of living organisms. The energy needed to perform these conversions they imagined to be supplied by volcanism, radioactive decay, ultraviolet radiation, meteoric impacts, or by lightning strokes.

The earth is thought to be approximately 4.6 billion years old. At the time when Calvin and his co-workers were performing their experiments, the earth's primitive atmosphere was believed to have consisted primarily of hydrogen, water, ammonia, methane, and carbon monoxide, with a little carbon dioxide. A large quantity of hydrogen was believed to have been initially present in the primitive atmosphere, but it was thought to have been lost gradually over a period of time because the earth's gravitational attraction is too weak to effectively hold such a light and rapidly-moving molecule. However, Calvin and his group assumed sufficient hydrogen to be present to act as a reducing agent. In their 1950 experiments they subjected a mixture of hydrogen and carbon dioxide, with a catalytic amount of Fe^{2+}, to bombardment by fast particles from the Berkeley cyclotron. Their experiments resulted in a good yield of formic acid and a moderate yield of formaldehyde. (The fast particles from the cyclotron were designed to simulate an energy input from radioactive decay on the primitive earth.)

Two years later, Stanley Miller, working in the laboratory of Harold Urey at the University of Chicago, performed a much more refined experiment of the same type. In Miller's experiment, a mixture of the gases methane, ammonia, water and hydrogen was subjected to an energy input from an electric spark. Miller's apparatus was designed so that the gases were continuously circulated, passing first through the spark chamber, then

---

[4] It is now believed that the main constituents of the primordial atmosphere were carbon dioxide, water, nitrogen, and a little methane.

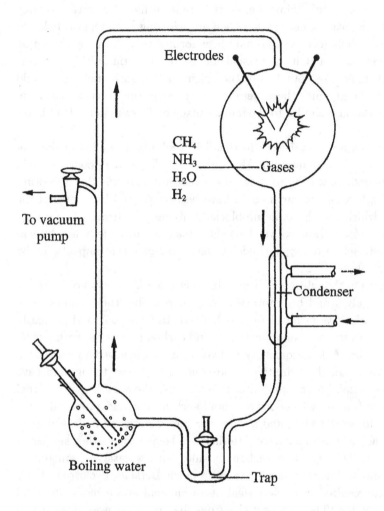

Fig. 3.4   Miller's apparatus.

through a water trap which removed the non-volatile water soluble products, and then back again through the spark chamber, and so on. The resulting products are shown as a function of time in Figure 3.5.

The Miller-Urey experiment produced many of the building-blocks of living organisms, including glycine, glycolic acid, sarcosine, alanine, lactic acid, N-methylalanine, β-alanine, succinic acid, aspartic acid, glutamic acid, iminodiacetic acid, iminoacetic-propionic acid, formic acid, acetic

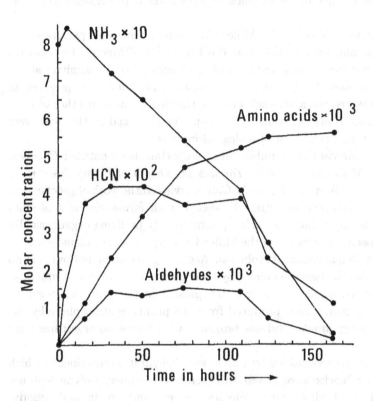

Fig. 3.5 Products as a function of time in the Miller-Urey experiment.

acid, propionic acid, urea and N-methyl urea[5]. Another major product was hydrogen cyanide, whose importance as an energy source in chemical evolution was later emphasized by Calvin.

The Miller-Urey experiment was repeated and extended by the Ceylonese-American biochemist Cyril Ponnamperuma and by the American expert in planetary atmospheres, Carl Sagan. They showed that when phosphorus is made available, then in addition to amino acids, the Miller-Urey experiment produces not only nucleic acids of the type that join together to form DNA, but also the energy-rich molecule ATP (adenosine triphosphate). ATP is extremely important in biochemistry, since it is a

---

[5] The chemical reaction that led to the formation of the amino acids that Miller observed was undoubtedly the Strecker synthesis: $HCN + NH_3 + RC=O + H_2O \rightarrow RC(NH_2)COOH$.

universal fuel which drives chemical reactions inside present-day living organisms.

Further variations on the Miller-Urey experiment were performed by Sydney Fox and his co-workers at the University of Miami. Fox and his group showed that amino acids can be synthesized from a primitive atmosphere by means of a thermal energy input, and that in the presence of phosphate esters, the amino acids can be thermally joined together to form polypeptides. However, some of the peptides produced in this way were cross linked, and hence not of biological interest.

In 1969, Melvin Calvin published an important book entitled *Chemical Evolution; Molecular Evolution Towards the Origin of Living Systems on Earth and Elsewhere*. In this book, Calvin reviewed the work of geochemists showing the presence in extremely ancient rock formations of molecules which we usually think of as being produced only by living organisms. He then discussed experiments of the Miller-Urey type — experiments simulating the first step in chemical evolution. According to Calvin, not only amino acids but also the bases adenine, thymine, guanine, cytosine and uracil, as well as various sugars, were probably present in the primitive ocean in moderate concentrations, produced from the primitive atmosphere by the available energy inputs, and not broken down because no organisms were present.

The next steps visualized by Calvin were dehydration reactions in which the building blocks were linked together into peptides, polynucleotides, lipids and porphyrins. Such dehydration reactions are in a thermodynamically uphill direction. In modern organisms, they are driven by a universally-used energy source, the high-energy phosphate bond of adenosine triphosphate (ATP). Searching for a substance present in the primitive ocean which could have driven the dehydrations, Calvin and his coworkers experimented with hydrogen cyanide ($HC\equiv N$), and from the results of these experiments they concluded that the energy stored in the carbon-nitrogen triple bond of $HC\equiv N$ could indeed have driven the dehydration reactions necessary for polymerization of the fundamental building blocks. However, later work made it seem improbable that peptides could be produced from cyanide mixtures.

In Chemical Evolution, Calvin introduced the concept of autocatalysis as a mechanism for molecular selection, closely analogous to natural selection in biological evolution. Calvin proposed that there were a few molecules in the ancient oceans which could catalyze the breakdown of the energy-rich molecules present into simpler products. According to Calvin's hypothesis, in a very few of these reactions, the reaction itself produced more of the catalyst. In other words, in certain cases the catalyst not only broke down the energy-rich molecules into simpler products but also catalyzed their own synthesis. These autocatalysts, according to Calvin, were the first systems which might possibly be regarded as living organisms. They not only "ate" the energy-rich molecules but they also reproduced, i.e., they catalyzed the synthesis of molecules identical with themselves.

Autocatalysis leads to a sort of molecular natural selection, in which the precursor molecules and the energy-rich molecules play the role of "food", and the autocatalytic systems compete with each other for the food supply. In Calvin's picture of molecular evolution, the most efficient autocatalytic systems won this competition in a completely Darwinian way. These more efficient autocatalysts reproduced faster and competed more successfully for precursors and for energy-rich molecules. Any random change in the direction of greater efficiency was propagated by natural selection.

What were these early autocatalytic systems, the forerunners of life? Calvin proposed several independent lines of chemical evolution, which later, he argued, joined forces. He visualized the polynucleotides, the polypeptides, and the metallo-porphyrins as originally having independent lines of chemical evolution. Later, he argued, an accidental union of these independent autocatalysts showed itself to be a still more efficient autocatalytic system. He pointed out in his book that "autocatalysis" is perhaps too strong a word. One should perhaps speak instead of "reflexive catalysis" , where a molecule does not necessarily catalyze the synthesis of itself, but perhaps only the synthesis of a precursor. Like autocatalysis, reflexive catalysis is capable of exhibiting Darwinian selectivity.

The theoretical biologist, Stuart Kauffman, working at the Santa Fe Institute, has constructed computer models for the way in which the components of complex systems of reflexive catalysts may have been linked together. Kauffman's models exhibit a surprising tendency to produce orderly behavior even when the links are randomly programmed.

In 1967 and 1968, C. Woese, F.H.C. Crick and L.E. Orgel proposed that there may have been a period of chemical evolution involving RNA alone, prior to the era when DNA, RNA and proteins joined together to form complex self-reproducing systems. In the early 1980's, this picture of an "RNA world" was strengthened by the discovery (by Thomas R. Cech and Sydney Altman) of RNA molecules which have catalytic activity.

Today experiments aimed at throwing light on chemical evolution towards the origin of life are being performed in the laboratory of the Nobel

Laureate geneticist Jack Sjostak at Harvard Medical School. The laboratory is trying to build a synthetic cellular system that undergoes Darwinian evolution.

In connection with autocatalytic systems, it is interesting to think of the polymerase chain reaction, which we discussed above. The target segment of DNA and the polymerase together form an autocatalytic system. The "food" molecules are the individual nucleotides in the solution. In the PCR system, a segment of DNA reproduces itself with an extremely high degree of fidelity. One can perhaps ask whether systems like the PCR system can have been among the forerunners of living organisms. The cyclic changes of temperature needed for the process could have been supplied by the cycling of water through a hydrothermal system. There is indeed evidence that hot springs and undersea hydrothermal vents may have played an important role in chemical evolution towards the origin of life. We will discuss this evidence in the next section.

Throughout this discussion of theories of chemical evolution, and the experiments which have been done to support these theories, energy has played a central role. None of the transformations discussed above could have taken place without an energy source, or to be more precise, they could not have taken place without a source of free energy. In Chapter 4 we will discuss in detail the reason why free energy plays a central role, not only in the origin of life but also in life's continuation. We will see that there is a connection between free energy and information, and that information-containing free energy is needed to produce the high degree of order which is characteristic of life.

Molecular evidence establishing family trees in evolution

Starting in the 1970's, the powerful sequencing techniques developed by Sanger and others began to be used to establish evolutionary trees. The evolutionary closeness or distance of two organisms could be estimated from the degree of similarity of the amino acid sequences of their proteins, and also by comparing the base sequences of their DNA and RNA. One of the first studies of this kind was made by R.E. Dickerson and his coworkers, who studied the amino acid sequences in Cytochrome C, a protein of very ancient origin which is involved in the "electron transfer chain" of respiratory metabolism. Some of the results of Dickerson's studies are shown in Figure 3.6.

Comparison of the base sequences of RNA and DNA from various species

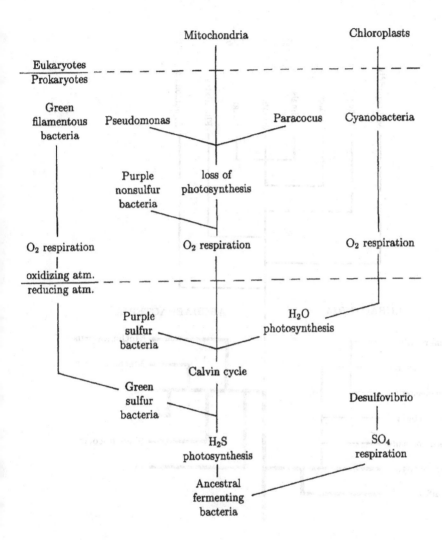

Fig. 3.6   Evolutionary relationships established by Dickerson and coworkers by comparing the amino acid sequences of Cytochrome C from various species.

proved to be even more powerful tool for establishing evolutionary relationships. Figure 3.7 shows the universal phylogenetic tree established in this way by Iwabe, Woese and their coworkers.[6] In Figure 3.7, all presently

---

[6] "Phylogeny" means "the evolutionary development of a species". "Ontogeny" means "the growth and development an individual, through various stages, for example, from fertilized egg to embryo, and so on". Ernst Haeckel, a 19th century follower of Darwin,

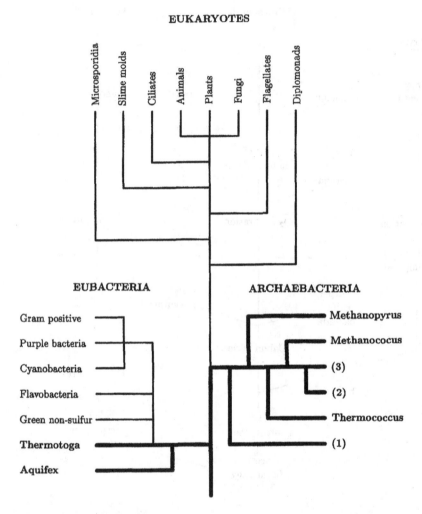

Fig. 3.7   This figure shows the universal phylogenetic tree, established by the work of Woese, Iwabe et al. Hyperthermophiles are indicated by bold lines and by bold type.

living organisms are divided into three main kingdoms, Eukaryotes, Eubacteria, and Archaebacteria. Carl Woese, who proposed this classification on the basis of comparative sequencing, wished to call the three kingdoms "Eucarya, Bacteria and Archaea". However, the most widely accepted terms

observed that, in many cases, "ontogeny recapitulates phylogeny".

are the ones shown in capital letters on the figure. Before the comparative RNA sequencing work, which was performed on the ribosomes of various species, it had not been realized that there are two types of bacteria, so markedly different from each other that they must be classified as belonging to separate kingdoms. One example of the difference between archaebacteria and eubacteria is that the former have cell membranes which contain ether lipids, while the latter have ester lipids in their cell membranes. Of the three kingdoms, the eubacteria and the archaebacteria are "prokaryotes", that is to say, they are unicellular organisms having no cell nucleus. Most of the eukaryotes, whose cells contain a nucleus, are also unicellular, the exceptions being plants, fungi and animals.

One of the most interesting features of the phylogenetic tree shown in Figure 3.7 is that the deepest branches — the organisms with shortest pedigrees — are all hyperthermophiles, i.e., they live in extremely hot environments such as hot springs or undersea hydrothermal vents. The shortest branches represent the most extreme hyperthermophiles. The group of archaebacteria indicated by (1) in the figure includes **Thermofilum, Thermoproteus, Pyrobaculum, Pyrodictium, Desulfurococcus**, and **Sulfolobus** — all hypothermophiles[7]. Among the eubacteria, the two shortest branches, Aquifex and Thermatoga are both hyperthermophiles[8]

The phylogenetic evidence for the existence of hyperthermophiles at a very early stage of evolution lends support to a proposal put forward in 1988 by the German biochemist Günter Wächterhäuser. He proposed that the reaction for pyrite formation,

$$FeS + H_2S \rightarrow FeS_2 + 2H + +2e^-$$

which takes place spontaneously at high temperatures, supplied the energy needed to drive the first stages of chemical evolution towards the origin of life. Wächterhäuser pointed out that the surface of the mineral pyrite (FeS_2) is positively charged, and he proposed that, since the immediate products of carbon-dioxide fixation are negatively charged, they would be attracted to the pyrite surface. Thus, in Wächterhäuser's model, pyrite formation not only supplied the reducing agent needed for carbon-dioxide

---

[7] Group (2) in Figure 3.7 includes **Methanothermus**, which is hyperthermophilic, and Methanobacterium, which is not. Group (3) includes **Archaeoglobus**, which is hyperthermophilic, and Halococcus, Halobacterium, Methanoplanus, Methanospirilum, and Methanosarcina, which are not.

[8] Thermophiles are a subset of the larger group of extremophiles.

fixation, but also the pyrite surface aided the process. Wächterhäuser further proposed an archaic autocatylitic carbon-dioxide fixation cycle, which he visualized as resembling the reductive citric acid cycle found in present-day organisms, but with all reducing agents replaced by FeS + H_2S, with thioester activation replaced by thioacid activation, and carbonyl groups replaced by thioenol groups. The interested reader can find the details of Wächterhäuser's proposals in his papers, which are listed at the end of this chapter.

A similar picture of the origin of life was proposed by Michael J. Russell and Alan J. Hall in 1997. In this picture "...(i) life emerged as hot, reduced, alkaline, sulphide-bearing submarine seepage waters interfaced with colder, more oxidized, more acid, Fe^{2+} >>Fe^{3+}-bearing water at deep (ca. 4km) floors of the Hadean ocean ca. 4 Gyr ago; (ii) the difference in acidity, temperature and redox potential provided a gradient of pH (ca. 4 units), temperature (ca. 60°C) and redox potential (ca. 500 mV) at the interface of those waters that was sustainable over geological time-scales, providing the continuity of conditions conducive to organic chemical reactions needed for the origin of life..." [9]. Russell, Hall and their coworkers also emphasize the role that may have been played by spontaneously-formed 3-dimensional mineral chambers (bubbles). They visualize these as having prevented the reacting molecules from diffusing away, thus maintaining high concentrations.

Table 3.2 shows the energy-yielding reactions which drive the metabolisms of some organisms which are of very ancient evolutionary origin. All the reactions shown in the table make use of H_2, which could have been supplied by pyrite formation at the time when the organisms evolved. All these organisms are lithoautotrophic, a word which requires some explanation: A heterotrophic organism is one which lives by ingesting energy-rich organic molecules which are present in its environment. By contrast, an autotrophic organism ingests only inorganic molecules. The lithoautotrophs use energy from these inorganic molecules, while the metabolisms of photoautotrophs are driven by energy from sunlight.

Evidence from layered rock formations called "stromatolites", produced by colonies of photosynthetic bacteria, show that photoautotrophs (or phototrophs) appeared on earth at least 3.5 billion years ago. The geological

---

[9]See W. Martin and M.J. Russell, *On the origins of cells: a hypothesis for the evolutionary transitions from abiotic geochemistry to chemoautotrophic prokaryotes, and from prokaryotes to nucleated cells*, Philos. Trans. R. Soc. Lond. B Biol. Sci., **358**, 59-85, (2003).

Table 3.2: **Energy-yielding reactions of some lithoautotrophic hyperthermophiles. (After K.O. Setter)**

Energy-yielding reaction	Genera
$4H_2+CO_2 \rightarrow CH_4+2H_2O$	Methanopyrus, Methanothermus, Methanococcus
$H_2+S° \rightarrow H_2S$	Pyrodictium, Thermoproteus, Pyrobaculum, Acidianus, Stygiolobus
$4H_2+H_2SO_4 \rightarrow H_2S+4H_2O$	Archaeoglobus

record also supplies approximate dates for other events in evolution. For example, the date at which molecular oxygen started to become abundant in the earth's atmosphere is believed to have been 2.0 billion years ago, with equilibrium finally being established 1.5 billion years in the past. Multicellular organisms appeared very late on the evolutionary and geological time-scale — only 600 million years ago. By collecting such evidence, the Belgian cytologist Christian de Duve has constructed the phylogenetic tree shown in Figure 3.8, showing branching as a function of time. One very interesting feature of this tree is the arrow indicating the transfer of "endosymbionts" from the eubacteria to the eukaryotes. In the next section, we will look in more detail at this important event, which took place about 1.8 billion years ago.

Symbiosis

The word "symbiosis" is derived from Greek roots meaning "living together". It was coined in 1877 by the German botanist Albert Bernard Frank. By that date, it had become clear that lichens are composite organisms involving a fungus and an alga; but there was controversy concerning whether the relationship was a parasitic one. Was the alga held captive and exploited by the fungus? Or did the alga and the fungus help each other,

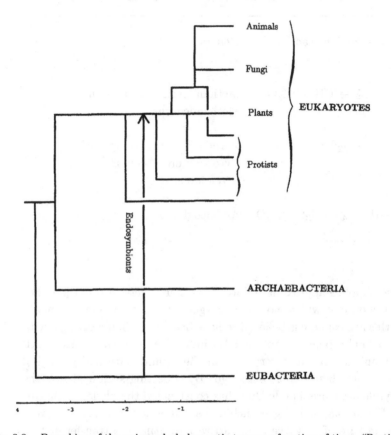

Fig. 3.8   Branching of the universal phylogenetic tree as a function of time. "Protists" are unicellular eukaryotes.

the former performing photosynthesis, and the latter leeching minerals from the lichen's environment? In introducing the word "symbiosis" (in German, "Symbiotismus"), Frank remarked that "We must bring all the cases where two different species live on or in one another under a comprehensive concept which does not consider the role which the two individuals play but is based on the mere coexistence, and for which the term symbiosis is to be recommended". Thus the concept of symbiosis, as defined by Frank, included all intimate relationships between two or more species, including parasitism at one extreme and "mutualism" at the other. However, as the word is used today, it usually refers to relationships which are mutually beneficial.

Charles Darwin himself had been acutely aware of close and mutually beneficial relationships between organisms of different species. For example, in his work on the fertilization of flowers, he had demonstrated the way in which insects and plants can become exquisitely adapted to each other's needs. However, T.H. Huxley, "Darwin's bulldog", emphasized competition as the predominant force in evolution. "The animal world is on about the same level as a gladiator's show", Huxley wrote in 1888, "The creatures are fairly well treated and set to fight — whereby the strongest, the swiftest and the cunningest live to fight another day. The spectator has no need to turn his thumbs down, as no quarter is given". The view of nature as a sort of "gladiator's contest" dominated the mainstream of evolutionary thought far into the 20th century; but there was also a growing body of opinion which held that symbiosis could be an extremely important mechanism for the generation of new species.

Among the examples of symbiosis studied by Frank were the nitrogen-fixing bacteria living in nodules on the roots of legumes, and the mycorrhizal fungi which live on the roots of forest trees such as oaks, beech and conifers. Frank believed that the mycorrhizal fungi aid in the absorption of nutrients. He distinguished between "ectotrophic" fungi, which form sheaths around the root fibers, and "endotrophic" fungi, which penetrate the root cells. Other examples of symbiosis studied in the 19th century included borderline cases between plants and animals, for ex- ample, paramecia, sponges, hydra, planarian worms and sea anemones, all of which frequently contain green bodies capable of performing photosynthesis.

Writing in 1897, the American lichenologist Albert Schneider prophesied that "future studies may demonstrate that.., plasmic bodies (within the eukaryote cell), such as chlorophyll granules, leucoplastids, chromoplastids, chromosomes, centrosomes, nucleoli, etc., are perhaps symbionts comparable to those in less highly specialized symbiosis. Reinke expresses the opinion that it is not wholly unreasonable to suppose that some highly skilled scientist of the future may succeed in cultivating chlorophyll-bodies in artificial media".

Nineteenth century cytologists such as Robert Altman, Andreas Schimper and A. Benda focused attention on the chlorophyll-bodies of plants, which Schimper named chloroplasts, and on another type of subcellular granule, present in large numbers in all plant and animal cells, which Benda named mitochondria, deriving the name from the Greek roots mitos (thread) and chrondos (granule). They observed that these bodies seemed to reproduce themselves within the cell in very much the manner that might be expected if they were independent organisms. Schimper suggested that chloroplasts are symbionts, and that green plants owe their origin to a union of a colorless unicellular organism with a smaller chlorophyll-containing species.

The role of symbiosis in evolution continued to be debated in the 20th century. Mitochondria were shown to be centers of respiratory metabolism; and it was discovered that both mitochondria and chloroplasts contain their own DNA. However, opponents of their symbiotic origin pointed out that mitochondria alone cannot synthesize all their own proteins: Some mitochondrial proteins require information from nuclear DNA. The debate was finally settled in the 1970's, when comparative sequencing of ribosomal RNA in the laboratories of Carl Woese, W. Ford Doolittle and Michael Gray showed conclusively that both chloroplasts and mitochondria were originally endosymbionts. The ribosomal RNA sequences showed that chloroplasts had their evolutionary root in the cyanobacteria, a species of eubacteria, while mitochondria were traced to a group of eubacteria called the alpha-proteobacteria. Thus the evolutionary arrow leading from the eubacteria to the eukaryotes can today be drawn with confidence, as in Figure 3.8.

Cyanobacteria are bluish photosynthetic bacteria which often become linked to one another so as to form long chains. They can be found today growing in large colonies on seacoasts in many parts of the world, for example in Baja California on the Mexican coast. The top layer of such colonies consists of the phototrophic cyanobacteria, while the organisms in underlying layers are heterotrophs living off the decaying remains of the cyanobacteria. In the course of time, these layered colonies can become fossilized, and they are the source of the layered rock formations called stromatolites (discussed above). Geological dating of ancient stromatolites has shown that cyanobacteria must have originated at least 3.5 billion years ago.

Cyanobacteria contain two photosystems, each making use of a different type of chlorophyll. Photosystem I, which is thought to have evolved first, uses the energy of light to draw electrons from inorganic compounds, and sometimes also from organic compounds (but never from water). Photosystem II, which evolved later, draws electrons from water. Hydrogen derived from the water is used to produce organic compounds from carbon dioxide, and molecular oxygen is released into the atmosphere. Photosystem II never appears alone. In all organisms which possess it, Photosystem II is coupled to Photosystem I, and together the two systems raise electrons to energy levels that are high enough to drive all the processes of metabolism. Dating of ancient stromatolites makes it probable that cyanobacteria began to release molecular oxygen into the earth's atmosphere at least 3.5 billion years ago; yet from other geological evidence we know that it was only 2 billion years ago that the concentration of molecular oxygen began to rise, equilibrium being reached 1.5 billion years ago. It is believed that ferrous iron, which at one time was very abundant, initially absorbed the photosynthetically produced oxygen. This resulted in the time-lag, as well as the ferrous-ferric mixture of iron which is found in the mineral magnetite.

When the concentrations of molecular oxygen began to rise in earnest, most of the unicellular microorganisms living at the time found themselves in deep trouble, faced with extinction, because for them, oxygen was a deadly poison; and very many species undoubtedly perished. However, some of the archaebacteria retreated to isolated anaerobic niches where we find them today, while others found ways of detoxifying the poisonous oxygen. Among the eubacteria, the ancestors of the alpha-proteobacteria were particularly good at dealing with oxygen and even turning it to advantage: They developed the biochemical machinery needed for respiratory metabolism.

Meanwhile, during the period between 3.5 and 2.0 billion years before the present, an extremely important evolutionary development had taken place: Branching from the archaebacteria, a line of large[10] heterotrophic unicellular organisms had evolved. They lacked rigid cell walls, and they could surround smaller organisms with their flexible outer membrane, drawing the victims into their interiors to be digested. These new heterotrophs were the ancestors of present-day eukaryotes, and thus they were the ancestors of all multicellular organisms.

Not only are the cells of present-day eukaryotes very much larger than the cells of archaebacteria and eubacteria; their complexity is also astonishing. Every eukaryote cell contains numerous intricate structures: a nucleus, cytoskeleton, Golgi apparatus, endoplasmic reticulum, mitochondria, peroxisomes, chromosomes, the complex structures needed for mitotic cell division, and so on. Furthermore, the genomes of eukaryotes contain very much more information than those of prokaryotes. How did this huge and relatively sudden increase in complexity and information content take place?

---

[10] Not large in an absolute sense, but large in relation to the prokaryotes.

According to a growing body of opinion, symbiosis played an important role in this development.

The ancestors of the eukaryotes were in the habit of drawing the smaller prokaryotes into their interiors to be digested. It seems likely that in a few cases the swallowed prokaryotes resisted digestion, multiplied within the host, were transmitted to future generations when the host divided, and conferred an evolutionary advantage, so that the result was a symbiotic relationship. In particular, both mitochondria and chloroplasts have definitely been proved to have originated as endosymbionts. It is easy to understand how the photosynthetic abilities of the chloroplasts (derived from cyanobacteria) could have conferred an advantage to their hosts, and how mitochondria (derived from alpha-proteobacteria) could have helped their hosts to survive the oxygen crisis. The symbiotic origin of other subcellular organelles is less well understood and is currently under intense investigation.

If we stretch the definition of symbiosis a little, we can make the concept include cooperative relationships between organisms of the same species. For example, cyanobacteria join together to form long chains, and they live together in large colonies which later turn into stromatolites. Also, some eubacteria have a mechanism for sensing how many of their species are present, so that they know, like a wolf pack, when it is prudent to attack a larger organism. This mechanism, called "quorum sensing", has recently attracted much attention among medical researchers.

The cooperative behavior of a genus of unicellular eukaryotes called slime molds is particularly interesting because it gives us a glimpse of how multicellular organisms may have originated. The name of the slime molds is misleading, since they are not fungi, but heterotrophic protists similar to amoebae. Under ordinary circumstances, the individual cells wander about independently searching for food, which they draw into their interiors and digest, a process called "phagocytosis". However, when food is scarce, they send out a chemical signal of distress. Researchers have analyzed the molecule which expresses slime mold unhappiness, and they have found it to be cyclic adenosine monophosphate (cAMP). At this signal, the cells congregate and the mass of cells begins to crawl, leaving a slimy trail. At it crawls, the community of cells gradually develops into a tall stalk, surmounted by a sphere — the "fruiting body". Inside the sphere, spores are produced by a sexual process. If a small animal, for example a mouse, passes by, the spores may adhere to its coat; and in this way they may be transported to another part of the forest where food is more plentiful.

Thus slime molds represent a sort of missing link between unicellular and multicellular or organisms. Normally the cells behave as individualists, wandering about independently, but when challenged by a shortage of food, the slime mold cells join together into an entity which closely resembles a multicellular organism. The cells even seem to exhibit altruism, since those forming the stalk have little chance of survival, and yet they are willing to perform their duty, holding up the sphere at the top so that the spores will survive and carry the genes of the community into the future. We should especially notice the fact that the cooperative behavior of the slime mold cells is coordinated by chemical signals.

Sponges are also close to the borderline which separates unicellular eukaryotes (protists) from multicellular organisms, but they are just on the other side of the border. Normally the sponge cells live together in a multicellular community, filtering food from water. However, if a living sponge is forced through a very fine cloth, it is possible to separate the cells from each other. The sponge cells can live independently for some time; but if many of them are left near to one another, they gradually join together and form themselves into a new sponge, guided by chemical signals. In a refinement of this experiment, one can take two living sponges of different species, separate the cells by passing the sponges through a fine cloth, and afterwards mix all the separated cells together. What happens next is amazing: The two types of sponge cells sort themselves out and become organized once more into two sponges — one of each species.

Slime molds and sponges hint at the genesis of multicellular organisms, whose evolution began approximately 600 million years ago. Looking at the slime molds and sponges, we can imagine how it happened. Some unicellular organisms must have experienced an enhanced probability of survival when they lived as colonies. Cooperative behavior and division of labor within the colonies were rewarded by the forces of natural selection, with the selective force acting on the entire colony of cells, rather than on the individual cell. This resulted in the formation of cellular societies and the evolution of mechanisms for cell differentiation. The division of labor within cellular societies (i.e., differentiation) came to be coordinated by chemical signals which affected the transcription of genetic information and the synthesis of proteins. Each cell within a society of cells possessed the entire genome characteristic of the colony, but once a cell had been assigned its specific role in the economy of the society, part of the information became blocked — that is, it was not expressed in the function of that particular cell. As multicellular organisms evolved, the chemical language of intercellular communication became very much more complex and refined. We will discuss the language of intercellular communication in more detail in a later section.

Geneticists have become increasingly aware that symbiosis has probably played a major role in the evolution of multicellular organisms. We mentioned above that, by means of genetic engineering techniques, transgenic plants and animals can be produced. In these chimeras, genetic material from a foreign species is incorporated into the chromosomes, so that it is inherited in a stable, Mendelian fashion. J.A. Shapiro, one of whose articles is referenced at the end of this chapter, believes that this process also occurs in nature, so that the conventional picture of evolutionary family trees needs to be corrected. Shapiro believes that instead of evolutionary trees, we should perhaps think of webs or networks.

For example, it is tempting to guess that symbiosis may have played a role in the development of the visual system of vertebrates. One of the archaebacteria, the purple halobacterium halobium (recently renamed halobacterium salinarum), is able to perform photosynthesis by means of a protein called bacterial rhodopsin, which transports hydrogen ions across the bacterial membrane. This protein is a near chemical relative of rhodopsin, which combines with a carotinoid to form the "visual purple" used in the vertebrate eye. It is tempting to think that the close similarity of the two molecules is not just a coincidence, and that vertebrate vision originated in a symbiotic relationship between the photosynthetic halobacterium and an aquatic ancestor of the vertebrates, the host being able to sense when the halobacterium was exposed to light and therefore transporting hydrogen ions across its cell membrane.

In this chapter, we have looked at the flow of energy and information in the origin and evolution of life on earth. We have seen how energy-rich molecules were needed to drive the first steps in the origin of life, and how during the evolutionary process, information was preserved, transmitted, and shared between increasingly complex organisms, the whole process being driven by an input of energy. In the next chapter, we will look closely at the relationships between energy and information.

Suggestions for further reading

(1) H. Lodish, A. Berk, S.L. Zipursky, P. Matsudaira, D. Baltimore, and J. Darnell, *Molecular Cell Biology, 4th Edition*, W.H. Freeman, New York, (2000).

(2) Lily Kay, *Who Wrote the Book of Life? A History of the Genetic Code*, Stanford University Press, Stanford CA, (2000).

(3) Sahotra Sarkar (editor), *The Philosophy and History of Molecular Biology*, Kluwer Academic Publishers, Boston, (1996).

(4) James D. Watson et al. *Molecular Biology of the Gene, 4th Edition*, Benjamin-Cummings, (1988).

(5) J.S. Fruton, *Proteins, Enzymes, and Genes*, Yale University Press, New Haven, (1999).

(6) S.E. Lauria, *Life, the Unfinished Experiment*, Charles Scribner's Sons, New York (1973).

(7) A. Lwoff, *Biological Order*, MIT Press, Cambridge MA, (1962).

(8) James D. Watson, *The Double Helix*, Athenium, New York (1968).

(9) F. Crick, *The genetic code*, Scientific American, **202**, 66-74 (1962).

(10) F. Crick, *Central dogma of molecular biology*, Nature, **227**, 561-563 (1970).

(11) David Freifelder (editor), *Recombinant DNA, Readings from the Scientific American*, W.H. Freeman and Co. (1978).

(12) James D. Watson, John Tooze and David T. Kurtz, *Recombinant DNA, A Short Course*, W.H. Freeman, New York (1983).

(13) Richard Hutton, *Biorevolution, DNA and the Ethics of Man-Made Life*, The New American Library, New York (1968).

(14) Martin Ebon, *The Cloning of Man*, The New American Library, New York (1978).

(15) Sheldon Krimsky, *Genetic Alchemy: The Social History of the Recombinant DNA Controversy*, MIT Press, Cambridge Mass (1983).

(16) M. Lappe, *Germs That Won't Die*, Anchor/Doubleday, Garden City N.Y. (1982).

(17) M. Lappe, *Broken Code*, Sierra Club Books, San Francisco (1984).

(18) President's Commission for the Study of Ethical Problems in Medicine and Biomedical and Behavioral Research, *Splicing Life: The Social and Ethical Issues of Genetic Engineering with Human Beings*, U.S. Government Printing Office, Washington D.C. (1982).

(19) U.S. Congress, Office of Technology Assessment, *Impacts of Applied Genetics - Microorganisms, Plants and Animals*, U.S. Government Printing Office, Washington D.C. (1981).

(20) W.T. Reich (editor), *Encyclopedia of Bioethics*, The Free Press, New York (1978).

(21) Martin Brown (editor), *The Social Responsibility of the Scientist*, The Free Press, New York (1970).

(22) B. Zimmerman, *Biofuture*, Plenum Press, New York (1984).

(23) John Lear, *Recombinant DNA, The Untold Story*, Crown, New York (1978).

(24) B. Alberts, D. Bray, J. Lewis, M. Raff, K. Roberts and J.D. Watson, *Molecular Biology of the Cell*, Garland, New York (1983).

(25) C. Woese, *The Genetic Code; The Molecular Basis for Genetic Expression*, Harper and Row, New York, (1967).

(26) F.H.C. Crick, *The Origin of the Genetic Code*, J. Mol. Biol. **38**, 367-379 (1968).

(27) M.W. Nirenberg, *The genetic code: II*, Scientific American, **208**, 80-94 (1962).

(28) L.E. Orgel, *Evolution of the Genetic Apparatus*, J. Mol. Biol. **38**, 381-393 (1968).

(29) Melvin Calvin, *Chemical Evolution Towards the Origin of Life, on Earth and Elsewhere*, Oxford University Press (1969).

(30) R. Shapiro, *Origins: A Skeptic's Guide to the Origin of Life*, Summit Books, New York, (1986).

(31) J. William Schopf, *Earth's earliest biosphere: its origin and evolution*, Princeton University Press, Princeton, N.J., (1983).

(32) J. William Schopf (editor), *Major Events in the History of Life*, Jones and Bartlet, Boston, (1992).

(33) Robert Rosen, *Life itself: a comprehensive inquiry into the nature, origin and fabrication of life*, Colombia University Press, (1991).

(34) R.F. Gesteland, T.R Cech, and J.F. Atkins (editors), *The RNA World, 2nd Edition*, Cold Spring Harbor Laboratory Press, Cold Spring Harbor, New York, (1999).

(35) C. de Duve, Blueprint of a Cell, Niel Patterson Publishers, Burlington N.C., (1991).

(36) C. de Duve, *Vital Dust; Life as a Cosmic Imperative*, Basic Books, New York, (1995).

(37) F. Dyson, *Origins of Life*, Cambridge University Press, (1985).

(38) S.A. Kaufman, *Antichaos and adaption*, Scientific American, 265, 78-84, (1991).

(39) S.A. Kauffman, *The Origins of Order*, Oxford University Press, (1993).

(40) F.J. Varela and J.-P. Dupuy, *Understanding Origins: Contemporary Views on the Origin of Life, Mind and Society*, Kluwer, Dordrecht, (1992).

(41) Stefan Bengtson (editor) *Early Life on Earth; Nobel Symposium No.*

84, Colombia University Press, New York, (1994).

(42) Herrick Baltscheffsky, *Origin and Evolution of Biological Energy Conversion*, VCH Publishers, New York, (1996).

(43) J. Chilea-Flores, T. Owen and F. Raulin (editors), *First Steps in the Origin of Life in the Universe*, Kluwer, Dordrecht, (2001).

(44) R.E. Dickerson, Nature **283**, 210-212 (1980).

(45) R.E. Dickerson, Scientific American **242**, 136-153 (1980).

(46) C.R. Woese, *Archaebacteria*, Scientific American **244**, 98-122 (1981).

(47) N. Iwabe, K. Kuma, M. Hasegawa, S. Osawa and T. Miyata, *Evolutionary relationships of archaebacteria, eubacteria, and eukaryotes inferred phylogenetic trees of duplicated genes*, Proc. Nat. Acad. Sci. USA **86**, 9355-9359 (1989).

(48) C.R. Woese, O. Kundler, and M.L. Wheelis, *Towards a Natural System of Organisms: Proposal for the Domains Archaea, Bacteria and Eucarya*, Proc. Nat. Acad. Sci. USA **87**, 4576-4579 (1990).

(49) W. Ford Doolittle, Phylogenetic Classification and the Universal Tree, Science, **284**, (1999).

(50) G. Wächterhäuser, *Pyrite formation, the first energy source for life: A hypothesis*, Systematic and Applied Microbiology **10**, 207-210 (1988).

(51) G. Wächterhäuser, *Before enzymes and templates: Theory of surface metabolism*, Microbiological Reviews, **52**, 452-484 (1988).

(52) G. Wächterhäuser, *Evolution of the first metabolic cycles*, Proc. Nat. Acad. Sci. USA **87**, 200-204 (1990).

(53) G. Wächterhäuser, *Groundworks for an evolutionary biochemistry the iron-sulfur world*, Progress in Biophysics and Molecular Biology **58**, 85-210 (1992).

(54) M.J. Russell and A.J. Hall, *The emergence of life from iron monosulphide bubbles at a submarine hydrothermal redox and pH front* J. Geol. Soc. Lond. **154**, 377-402, (1997).

(55) L.H. Caporale (editor), *Molecular Strategies in Biological Evolution*, Ann. N.Y. Acad. Sci., May 18, (1999).

(56) W. Martin and M.J. Russell, *On the origins of cells: a hypothesis for the evolutionary transitions from abiotic geochemistry to chemoautotrophic prokaryotes, and from prokaryotes to nucleated cells*, Philos. Trans. R. Soc. Lond. B Biol. Sci., **358**, 59-85, (2003).

(57) Werner Arber, *Elements in Microbal Evolution*, J. Mol. Evol. **33**, 4 (1991).

(58) Michael Gray, *The Bacterial Ancestry of Plastids and Mitochondria*, BioScience, **33**, 693-699 (1983).

(59) Michael Grey, *The Endosymbiont Hypothesis Revisited*, International Review of Cytology, **141**, 233-257 (1992).

(60) Lynn Margulis and Dorian Sagan, *Microcosmos: Four Billion Years of Evolution from Our Microbal Ancestors*, Allan and Unwin, London, (1987).

(61) Lynn Margulis and Rene Fester, eds., *Symbiosis as as Source of Evolutionary Innovation: Speciation and Morphogenesis*, MIT Press, (1991).

(62) Charles Mann, *Lynn Margulis: Science's Unruly Earth Mother*, Science, **252**, 19 April, (1991).

(63) Jan Sapp, *Evolution by Association; A History of Symbiosis*, Oxford University Press, (1994).

(64) J.A. Shapiro, *Natural genetic engineering in evolution*, Genetics, **86**, 99-111 (1992).

(65) E.M. De Robertis et al., *Homeobox genes and the vertebrate body plan*, Scientific American, July, (1990).

(66) J.S. Schrum, T.F. Zhu and J.W. Szostak, *The origins of cellular life*, Cold Spring Harb. Perspect. Biol., May 19 (2010).

(67) I. Budin and J.W. Szostak, *Expanding Roles for Diverse Physical Phenomena During the Origin of Life*, Annu. Rev. Biophys., **39**, 245-263, (2010).

Chapter 4

STATISTICAL MECHANICS AND INFORMATION

The second law of thermodynamics

In this chapter, we discuss the origin and evolution of living organisms from the standpoint of thermodynamics, statistical mechanics and information theory. In particular, we discuss the work of Maxwell, Boltzmann, Gibbs, Szilard, and Shannon. Their research established the fact that free energy[1] contains information, and that it can thus be seen as the source of the order and complexity of living systems. The reader who prefers to avoid mathematics may jump quickly over the equations in this chapter without losing the thread of the argument, provided that he or she is willing to accept this conclusion.

Our starting point is the second law of thermodynamics, which was discovered by Nicolas Leonard Sadi Carnot (1796–1832) and elaborated by Rudolf Clausius (1822–1888) and William Thomson (later Lord Kelvin, 1824–1907). Carnot came from a family of distinguished French politicians and military men, but instead of following a political career, he studied engineering. In 1824, his only scientific publication appeared — a book with the title *Reflections on the Motive Power of Fire*. Although it was ignored for the first few years after its publication, this single book was enough to secure Carnot a place in history as the founder of the science of thermodynamics. In his book, Carnot introduced a scientific definition of work which we still use today — "weight lifted through a height"; in other words, force times distance.

At the time when Carnot was writing, much attention was being given to improving the efficiency of steam engines. Although James Watt's steam engines were far more efficient than previous models, they still could only

---

[1] I.e., energy from which work can be derived.

convert between 5 % and 7 % of the heat energy of their fuels into useful work. Carnot tried to calculate the theoretical maximum of the efficiency of steam engines, and he was able to show that an engine operating between the temperatures T_1 and T_2 could at most attain

$$\text{maximum efficiency} = \frac{T_1 - T_2}{T_1} \qquad (4.1)$$

Here T_1 is the temperature of the input steam, and T_2 is the temperature of the cooling water. Both these temperatures are absolute temperatures, i.e., temperatures proportional to the volume of a given quantity of gas at constant pressure.

Carnot died of cholera at the age of 36. Fifteen years after his death, the concept of absolute temperature was further clarified by Lord Kelvin (1824-1907), who also helped to bring Carnot's work to the attention of the scientific community.

Building on the work of Carnot, the German theoretical physicist Rudolph Clausius was able to deduce an extremely general law. He discovered that the ratio of the heat content of a closed system to its absolute temperature always increases in any process. He called this ratio the entropy of the system. In the notation of modern thermodynamics, the change in entropy dS when a small amount of heat dq is transferred to a system is given by

$$dS = \frac{dq}{dT} \qquad (4.2)$$

Let us imagine a closed system consisting of two parts, one at temperature T_1, and the other part at a lower temperature T_2. If a small amount of heat dq flows from the warmer part to the cooler one, the small resulting change in entropy of the total system will be

$$dS = \frac{dq}{T_1} - \frac{dq}{T_2} > 0 \qquad (4.3)$$

According to Clausius, since heat never flows spontaneously from a colder object to a warmer one, the entropy of a closed system always increases; that is to say, dS is always positive. As heat continues to flow from the warmer part of the system to the cooler part, the system's energy becomes less and less available for doing work. Finally, when the two parts have reached the same temperature, no work can be obtained. When the parts differed in temperature, a heat engine could in principle be run between them, making use of the temperature difference; but when the two parts have reached the same temperature, this possibility no longer exists. The law stating that the entropy of a closed system always increases is called the second law of thermodynamics.

Maxwell's demon

In England, the brilliant Scottish theoretical physicist, James Clerk Maxwell (1831-1879) invented a thought experiment which demonstrated that the second law of thermodynamics is statistical in nature and that there is a relationship between entropy and information. It should be mentioned that at the time when Clausius and Maxwell were living, not all scientists agreed about the nature of heat, but Maxwell, like Kelvin, believed heat to be due to the rapid motions of atoms or molecules. The more rapid the motion, the greater the temperature.

In a discussion of the ideas of Carnot and Clausius, Maxwell introduced a model system consisting of a gas-filled box divided into two parts by a wall; and in this wall, Maxwell imagined a small weightless door operated by a "demon". Initially, Maxwell let the temperature and pressure in both parts of the box be equal. However, he made his demon operate the door in such a way as to sort the gas particles: Whenever a rapidly-moving particle approaches from the left, Maxwell's demon opens the door; but when a slowly moving particle approaches from the left, the demon closes it. The demon has the opposite policy for particles approaching from the right, allowing the slow particles to pass, but turning back the fast ones. At the end of Maxwell's thought experiment, the particles are sorted, with the slow ones to the left of the barrier, and the fast ones to the right. Although initially, the temperature was uniform throughout the box, at the end a temperature difference has been established, the entropy of the total system is *decreased* and the second law of thermodynamics is violated.

In 1871, Maxwell expressed these ideas in the following words: "If we conceive of a being whose faculties are so sharpened that he can follow every molecule in its course, such a being, whose attributes are still finite as our own, would be able to do what is at present impossible to us. For we have seen that the molecules in a vessel full of air are moving at velocities by no means uniform... Now let us suppose that such a vessel full of air at a uniform temperature is divided into two portions, A and B, by a division in which there is a small hole, and that a being who can see individual molecules, opens and closes swifter molecules to pass from A to B, and only slower ones to pass from B to A. He will thus, without the expenditure of work, raise the temperature of B and lower that of A, in contradiction to the second law of thermodynamics". Of course Maxwell admitted that demons and weightless doors do not exist. However, he pointed out, one could certainly imagine a small hole in the partition between the two halves of the box. The sorting could happen by chance (although the probability of its happening decreases rapidly as the number of gas particles becomes large). By this argument, Maxwell demonstrated that the second law of thermodynamics is a statistical law.

An extremely interesting aspect of Maxwell's thought experiment is that his demon uses information to perform the sorting. The demon needs information about whether an approaching particle is fast or slow in order to know whether or not to open the door.

Finally, after the particles have been sorted, we can imagine that the partition is taken away so that the hot gas is mixed with the cold gas. During this mixing, the entropy of the system will increase, and information (about where to find fast particles and where to find slow ones) will be lost. Entropy is thus seen to be a measure of disorder or lack of information. To decrease the entropy of a system, and to increase its order, Maxwell's demon needs information. In the opposite process, the mixing process, where entropy increases and where disorder increases, information is lost.

Statistical mechanics

Besides inventing an interesting demon (and besides his monumental contributions to electromagnetic theory), Maxwell also helped to lay the foundations of statistical mechanics. In this enterprise, he was joined by the Austrian physicist Ludwig Boltzmann (1844–1906) and by an American, Josiah Willard Gibbs, whom we will discuss later. Maxwell and Boltzmann worked independently and reached similar conclusions, for which they share the credit. Like Maxwell, Boltzmann also interpreted an increase in entropy as an increase in disorder; and like Maxwell he was a firm believer in atomism at a time when this belief was by no means universal. For example, Ostwald and Mach, both important figure in German science at that time, refused to believe in the existence of atoms, in spite of the fact that Dalton's atomic ideas had proved to be so useful in chemistry. Towards the end of his life, Boltzmann suffered from periods of severe depression, perhaps because of attacks on his scientific work by Ostwald and others. In 1906, while on vacation near Trieste, he committed suicide — ironically, just a year before the French physicist J.B. Perrin produced irrefutable evidence of the existence of atoms.

Maxwell and Boltzmann made use of the concept of "phase space", a $6N$-dimensional space whose coordinates are the position and momentum coordinates of each of N particles. However, in discussing statistical mechanics we will use a more modern point of view, the point of view of quantum theory, according to which a system may be in one or another of a set of discrete states, $i = 1,2,3,...$ with energies ϵ_i. Let us consider a set of N identical, weakly-interacting systems; and let us denote the number of the systems which occupy a particular state by n_j, as shown in equation (4.4):

$$
\begin{array}{lllllll}
\text{State number} & 1 & 2 & 3 & ... & i & ... \\
\\
\text{Energy} & \epsilon_1, & \epsilon_2, & \epsilon_3, & ... & \epsilon_i, & ... \qquad (4.4)\\
\\
\text{Occupation number} & n_1, & n_2, & n_3, & ... & n_i, & ...
\end{array}
$$

the energy levels and their occupation numbers. This macrostate can be constructed in many ways, and each of these ways is called a "microstate": For example, the first of the N identical systems may be in state 1 and the second in state 2; or the reverse may be the case; and the two situations correspond to different microstates. From combinatorial analysis it is possible to show that the number of microstates corresponding to a given macrostate is given by:

$$W = \frac{N!}{n_1!n_2!n_3!...n_i!...} \qquad (4.5)$$

Boltzmann was able to show that the entropy S_N of the N identical systems is related to the quantity W by the equation

$$S_N = k \ln W \qquad (4.6)$$

where k is the constant which appears in the empirical law relating the pressure, volume and absolute temperature of an ideal gas;

$$PV = NkT \qquad (4.7)$$

This constant,

$$k = 1.38062 \times 10^{-23} \frac{\text{joule}}{\text{kelvin}} \qquad (4.8)$$

is called Boltzmann's constant in his honor. Boltzmann's famous equation relating entropy to missing information, equation (4.6), is engraved on his tombstone. A more detailed discussion of Boltzmann's statistical mechanics is given in Appendix 1.

Information theory — Shannon's formula

We have seen that Maxwell's demon needed information to sort gas particles and thus decrease entropy; and we have seen that when fast and slow particles are mixed so that entropy increases, information is lost. The relationship between entropy and lost or missing information was made quantitative by the Hungarian-American physicist Leo Szilard (1898–1964) and by the American mathematician Claude Shannon (1916–2001). In 1929, Szilard published an important article in Zeitschrift für Physik in which he analyzed Maxwell's demon. In this famous article, Szilard emphasized the connection between entropy and missing information. He was able to show that the entropy associated with a unit of information is $k \ln 2$, where k is Boltzmann's constant. We will discuss this relationship in more detail below.

Claude Shannon is usually considered to be the "father of information theory". Shannon graduated from the University of Michigan in 1936, and he later obtained a Ph.D. in mathematics from the Massachusetts Institute of Technology. He worked at the Bell Telephone Laboratories, and later became a professor at MIT. In 1949, motivated by the need of AT&T to quantify the amount of information that could be transmitted over a given line, Shannon published a pioneering study of information as applied to communication and computers. Shannon first examined the question of how many binary digits are needed to express a given integer Ω. In the decimal system we express an integer by telling how many 1's it contains, how many 10's, how many 100's, how many 1000's, and so on. Thus, for example, in the decimal system,

$$105 = 1 \times 10^2 + 0 \times 10^1 + 5 \times 10^0 \qquad (4.9)$$

Any integer greater than or equal to 100 but less than 1000 can be expressed with 3 decimal digits; any number greater than or equal to 1000 but less than 10,000 requires 4, and so on.

The natural language of computers is the binary system; and therefore Shannon asked himself how many binary digits are needed to express an integer of a given size. In the binary system, a number is specified by telling how many of the various powers of 2 it contains. Thus, the decimal integer 105, expressed in the binary system, is

$$1101001 \equiv 1 \times 2^6 + 1 \times 2^5 + 0 \times 2^4 + 1 \times 2^3 + 0 \times 2^2 + 0 \times 2^1 + 1 \times 2^0 \quad (4.10)$$

In the many early computers, numbers and commands were read in on punched paper tape, which could either have a hole in a given position, or else no hole. Shannon wished to know how long a strip of punched tape is needed to express a number of a given size — how many binary digits are needed? If the number happens to be an exact power of 2, then the answer is easy: To express the integer

$$\Omega = 2^n \tag{4.11}$$

one needs $n + 1$ binary digits. The first binary digit, which is 1, gives the highest power of 2, and the subsequent digits, all of them 0, specify that the lower powers of 2 are absent. Shannon introduced the word "bit" as an abbreviation of "binary digit". He generalized this result to integers which are not equal to exact powers of 2: Any integer greater than or equal to 2^{n-l}, but less than 2^n, requires n binary digits or "bits". In Shannon's theory, the bit became the unit of information. He defined the quantity of information needed to express an arbitrary integer Ω as

$$I = \log_2 \Omega \text{ bits} = \frac{\ln \Omega}{\ln 2} \text{ bits} = 1.442695 \ln \Omega \text{ bits} \tag{4.12}$$

or

$$I = K \ln \Omega \qquad K = 1.442695 \text{ bits} \tag{4.13}$$

Of course the information function I, as defined by equation (4.13), is in general not an integer, but if one wishes to find the exact number of binary digits required to express a given integer Ω, one can calculate I and round upwards[2].

Shannon went on to consider quantitatively the amount of information which is missing before we perform an experiment, the result of which we are unable to predict with certainty. (For example, the "experiment" might be flipping a coin or throwing a pair of dice.) Shannon first calculated the missing information, I_N, not for a single performance of the experiment but for N independent performances. Suppose that in a single performance, the probability that a particular result i will occur is given by P_i. If the experiment is performed N times, then as N becomes very large, the fraction of times that the result i occurs becomes more and more exactly equal to P_i. For example, if a coin is flipped N times, then as N becomes extremely large, the fraction of "heads" among the results becomes more and more nearly equal to $1/2$. However, some information is still missing because we still do not know the sequence of the results. Shannon was able to show from combinatorial analysis, that this missing information about the sequence of the results is given by

$$I_N = K \ln \Omega \tag{4.14}$$

---

[2] Similar considerations can also be found in the work of the statistician R.A. Fisher.

where

$$\Omega = \frac{N!}{n_1! n_2! n_3! ... n_i! ...} \qquad n_i \equiv NP_i \qquad (4.15)$$

or

$$I_N = K \ln \Omega = K \left[\ln(N!) - \sum_i \ln(n_i) \right] \qquad (4.16)$$

Shannon then used Sterling's approximation, $\ln(n_i!) \approx n_i(\ln n_i - 1)$, to rewrite (4.16) in the form

$$I_N = -KN \sum_i P_i \ln P_i \qquad (4.17)$$

Finally, dividing by N, he obtained the missing information prior to the performance of a single experiment:

$$I = -K \sum_i P_i \ln P_i \qquad (4.18)$$

For example, in the case of flipping a coin, Shannon's equation, (4.18), tells us that the missing information is

$$I = -K \left[\frac{1}{2} \ln \left(\frac{1}{2} \right) + \frac{1}{2} \ln \left(\frac{1}{2} \right) \right] = 1 \text{ bit} \qquad (4.19)$$

As a second example, we might think of an "experiment" where we write the letters of the English alphabet on 26 small pieces of paper. We then place them in a hat and draw out one at random. In this second example,

$$P_a = P_b = ... = P_z = \frac{1}{26} \qquad (4.20)$$

and from Shannon's equation we can calculate that before the experiment is performed, the missing information is

$$I = -K \left[\frac{1}{26} \ln \left(\frac{1}{26} \right) + \frac{1}{26} \ln \left(\frac{1}{26} \right) + ... \right] = 4.70 \text{ bits} \qquad (4.21)$$

If we had instead picked a letter at random out of an English book, the letters would not occur with equal probability. From a statistical analysis of the frequency of the letters, we would know in advance that

$$P_a = 0.078, \qquad P_b = 0.013, \qquad ... \qquad P_z = 0.001 \qquad (4.22)$$

Shannon's equation would then give us a slightly reduced value for the missing information:

$$I = -K \left[0.078 \ln 0.078 + 0.013 \ln 0.013 + ... \right] = 4.15 \text{ bits} \qquad (4.23)$$

Less information is missing when we know the frequencies of the letters, and Shannon's formula tells us exactly how much less information is missing.

When Shannon had been working on his equations for some time, he happened to visit the mathematician John von Neumann, who asked him how he was getting on with his theory of missing information. Shannon replied that the theory was in excellent shape, except that he needed a good name for "missing information". "Why don't you call it entropy?", von Neumann suggested. "In the first place, a mathematical development very much like yours already exists in Boltzmann's statistical mechanics, and in the second place, no one understands entropy very well, so in any discussion you will be in a position of advantage!" Like Leo Szilard, von Neumann was a Hungarian-American, and the two scientists were close friends. Thus von Neumann was very much aware of Szilard's paper on Maxwell's demon, with its analysis of the relationship between entropy and missing information. Shannon took von Neumann's advice, and used the word "entropy" in his pioneering paper on information theory. Missing information in general cases has come to be known as "Shannon entropy". But Shannon's ideas can also be applied to thermodynamics.

Entropy expressed as missing information

From the standpoint of information theory, the thermodynamic entropy S_N of an ensemble of N identical weakly-interacting systems in a given macrostate can be interpreted as the missing information which we would need in order to specify the state of each system, i.e., the microstate of the ensemble. Thus, thermodynamic information is defined to be the negative of thermodynamic entropy, i.e., the information that would be needed to specify the microstate of an ensemble in a given macrostate. Shannon's formula allows this missing information to be measured quantitatively. Applying Shannon's formula, equation (4.13), to the missing information in Boltzmann's problem we can identify W with Ω, S_N with I_N, and k with K:

$$W \to \Omega \qquad S_N \to I_N \qquad k \to K = \frac{1}{\ln 2} \text{ bits} \qquad (4.24)$$

so that

$$k \ln 2 = 1 \text{ bit} = 0.95697 \times 10^{-23} \frac{\text{joule}}{\text{kelvin}} \qquad (4.25)$$

and

$$k = 1.442695 \text{ bits} \qquad (4.26)$$

This implies that temperature has the dimension energy/bit:

$$1 \text{ degree Kelvin} = 0.95697 \times 10^{-23} \frac{\text{joule}}{\text{bit}} \qquad (4.27)$$

From this it follows that

$$1 \frac{\text{joule}}{\text{kelvin}} = 1.04496 \times 10^{23} \text{ bits} \qquad (4.28)$$

If we divide equation (4.28) by Avogadro's number we have

$$1 \frac{\text{joule}}{\text{kelvin mol}} = \frac{1.04496 \times 10^{23} \text{ bits/molecule}}{6.02217 \times 10^{23} \text{ molecules/mol}} = 0.17352 \frac{\text{bits}}{\text{molecule}} \qquad (4.29)$$

Figure 4.1 shows the experimentally-determined entropy of ammonia, NH_3, as a function of the temperature, measured in kelvins. It is usual to express entropy in joule/kelvin-mol; but it follows from equation (4.29) that entropy can also be expressed in bits/molecule, as is shown in the figure. Since

$$1 \text{ electron volt} = 1.6023 \times 10^{-19} \text{ joule} \qquad (4.30)$$

it also follows from equation (4.29) that

$$1 \frac{\text{electron volt}}{\text{kelvin}} = 1.6743 \times 10^4 \text{ bits} \qquad (4.31)$$

Thus, one electron-volt of energy, converted into heat at room temperature, $T = 298.15$ kelvin, will produce an entropy change (or thermodynamic information change) of

$$\frac{1 \text{ electron volt}}{298.15 \text{ kelvin}} = 56.157 \text{ bits} \qquad (4.32)$$

When a system is in thermodynamic equilibrium, its entropy has reached a maximum; but if it is not in equilibrium, its entropy has a lower value. For example, let us think of the case which was studied by Clausius when he introduced the concept of entropy: Clausius imagined an isolated system, divided into two parts, one of which has a temperature T_i, and the other a lower temperature, T_2. When heat is transferred from the hot part to the cold part, the entropy of the system increases; and when equilibrium is finally established at some uniform intermediate temperature, the entropy has reached a maximum. The difference in entropy between the initial state of Clausius' system and its final state is a measure of how far away from thermodynamic equilibrium it was initially. From the discussion given above, we can see that it is also possible to interpret this entropy difference as the system's initial content of thermodynamic information.

Similarly, when a photon from the sun reaches (for example) a drop of water on the earth, the initial entropy of the system consisting of the photon

Entropy

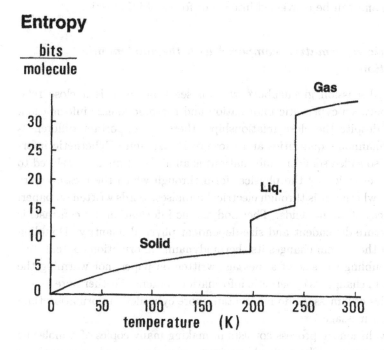

Fig. 4.1   This figure shows the entropy of ammonia as a function of temperature. It is usual to express entropy in joule/kelvin-mol, but it can also be expressed in bits/molecule.

plus the drop of water is smaller than at a later stage, when the photon's energy has been absorbed and shared among the water molecules, with a resulting very slight increase in the temperature of the water. This entropy difference can be interpreted as the quantity of thermodynamic information which was initially contained in the photon-drop system, but which was lost when the photon's free energy was degraded into heat. Equation (4.32) allows us to express this entropy difference in terms of bits. For example, if the photon energy is 2 electron-volts, and if the water drop is at a temperature of 298.15 degrees Kelvin, then $\Delta S = 112.31$ bits; and this amount of thermodynamic information is available in the initial state of the system. In our example, the information is lost; but if the photon had instead reached the leaf of a plant, part of its energy, instead of being immediately degraded, might have been stabilized in the form of high-energy chemical bonds. When a part of the photon energy is thus stabilized, not all of the thermodynamic information which it contains is lost; a part is conserved and can be converted into other forms of information.

Cybernetic information compared with thermodynamic information

From the discussion given above we can see that there is a close relationship between cybernetic information and thermodynamic information. However, despite the close relationships, there are important differences between Shannon's quantities and those of Boltzmann. Cybernetic information (also called semiotic information) is an abstract quantity related to messages, regardless of the physical form through which the messages are expressed, whether it is through electrical impulses, words written on paper, or sequences of amino acids. Thermodynamic information, by contrast, is a temperature-dependent and size-dependent physical quantity. Doubling the size of the system changes its thermodynamic information content; but neither doubling the size of a message written on paper, nor warming the message will change its cybernetic information content. Furthermore, many exact copies of a message do not contain more cybernetic information than the original message.

The evolutionary process consists in making many copies of a molecule or a larger system. The multiple copies then undergo random mutations; and after further copying, natural selection preserves those mutations that are favorable. It is thermodynamic information that drives the copying process, while the selected favorable mutations may be said to contain cybernetic information. The cybernetic information distilled in this process is always smaller than the quantity of thermodynamic information required to create it (both measured in bits) since all information must have a source.

The information content of Gibbs free energy

At the beginning of this chapter, we mentioned that the American physicist Josiah Willard Gibbs (1839-1903) made many contributions to thermodynamics and statistical mechanics. In 1863, Gibbs received from Yale the first Ph.D. in engineering granted in America, and after a period of further study in France and Germany, he became a professor of mathematical physics at Yale in 1871, a position which he held as long as he lived. During the period between 1876 and 1878, he published a series of papers in the *Transactions of the Connecticut Academy of Sciences*. In these papers, about 400 pages in all, Gibbs applied thermodynamics to chemical reactions. (The editors of the *Transactions of the Connecticut Academy of Sciences* did not really understand Gibbs' work, but, as they said later, "We knew Gibbs, and we took his papers on faith".)

Because the journal was an obscure one, and because Gibbs' work was so highly mathematical, it remained almost unknown to European scientists for a long period. However, in 1892 Gibbs' papers were translated into German by Ostwald, and in 1899 they were translated into French by Le Chatelier; and then the magnitude of Gibbs' contribution was finally recognized. One of his most important innovations was the definition of a quantity which we now call "Gibbs free energy". This quantity allows one to determine whether or not a chemical reaction will take place spontaneously.

Chemical reactions usually take place at constant pressure and constant temperature. If a reaction produces a gas as one of its products, the gas must push against the pressure of the earth's atmosphere to make a place for itself. In order to take into account the work done against external pressure in energy relationships, the German physiologist and physicist Hermann von Helmholtz introduced a quantity (which we now call heat content or enthalpy) defined by

$$H = U + PV \qquad (4.33)$$

where U is the internal energy of a system, P is the pressure, and V is the system's volume.

Gibbs went one step further than Helmholtz, and defined a quantity which would also take into account the fact that when a chemical reaction takes place, heat is exchanged with the surroundings. Gibbs defined his free energy by the relation

$$G = U + PV - TS \qquad (4.34)$$

or

$$G = H - TS \qquad (4.35)$$

where S is the entropy of a system, H is its enthalpy, and T is its temperature.

Gibbs' reason for introducing the quantity G is as follows: The second law of thermodynamics states that in any spontaneous process, the entropy of the universe increases. Gibbs invented a simple model of the universe, consisting of the system (which might, for example, be a beaker within which a chemical reaction takes place) in contact with a large thermal reservoir at constant temperature. The thermal reservoir could, for example, be a water bath so large that whatever happens in the chemical reaction, the temperature of the bath will remain essentially unaltered. In Gibbs' simplified model, the entropy change of the universe produced by the chemical reaction can be split into two components:

$$\Delta S_{universe} = \Delta S_{system} + \Delta S_{bath} \tag{4.36}$$

Now suppose that the reaction is endothermic (i.e., it absorbs heat). Then the reaction beaker will absorb an amount of heat ΔH_{system} from the bath, and the entropy change of the bath will be

$$\Delta S_{bath} = -\frac{\Delta H_{system}}{T} \tag{4.37}$$

Combining (4.36) and (4.37) with the condition requiring the entropy of the universe to increase, Gibbs obtained the relationship

$$\Delta S_{universe} = \Delta S_{system} - \frac{\Delta H_{system}}{T} > 0 \tag{4.38}$$

The same relationship also holds for exothermic reactions, where heat is transferred in the opposite direction. Combining equations (4.38) and (4.35) yields

$$\Delta G_{system} = -T\Delta S_{universe} < 0 \tag{4.39}$$

Thus, the Gibbs free energy for a system must decrease in any spontaneous chemical reaction or process which takes place at constant temperature and pressure. We can also see from equation (4.39) that Gibbs free energy is a measure of a system's content of thermodynamic information. If the available free energy is converted into heat, the quantity of thermodynamic information $\Delta S_{universe} = -\Delta G_{system}/T$ is lost, and we can deduce that in the initial state of the system, this quantity of information was available. Under some circumstances the available thermodynamic information can be partially conserved. In living organisms, chemical reactions are coupled together, and Gibbs free energy, with its content of thermodynamic information, can be transferred from one compound to another, and ultimately converted into other forms of information.

Measured values of the "Gibbs free energy of formation", ΔG_f°, are available for many molecules. To construct tables of these values, the change in Gibbs free energy is measured when the molecules are formed from their constituent elements. The most stable states of the elements at room temperature and atmospheric pressure are taken as zero points. For example, water in the gas phase has a Gibbs free energy of formation

$$\Delta G_f^\circ(H_2O) = -228.59 \, \frac{kJ}{mol} \tag{4.40}$$

This means that when the reaction

$$H_2(g) + \frac{1}{2}O_2(g) \rightarrow H_2O(g) \tag{4.41}$$

takes place under standard conditions, there is a change in Gibbs free energy of $\Delta G° = $ -228.59 kJ/mol [3]. The elements hydrogen and oxygen in their most stable states at room temperature and atmospheric pressure are taken as the zero points for Gibbs free energy of formation. Since $\Delta G°$ is negative for the reaction shown in equation (4.41), the reaction is spontaneous. In general, the change in Gibbs free energy in a chemical reaction is given by

$$\Delta G° = \sum_{products} \Delta G_f° - \sum_{reactants} \Delta G_f° \tag{4.42}$$

where $\Delta G_f°$ denotes the Gibbs free energy of formation.

As a second example, we can consider the reaction in which glucose is burned:

$$C_6H_{12}O_6(s) + 6O_2(g) \rightarrow 6CO_2(g) + 6H_2O(g) \qquad \Delta G° = -2870 \frac{kJ}{mol} \tag{4.43}$$

From equation (4.29) it follows that in this reaction,

$$-\frac{\Delta G°}{T} = 1670 \frac{bits}{molecule} \tag{4.44}$$

If the glucose is simply burned, this amount of information is lost; but in a living organism, the oxidation of glucose is usually coupled with other reactions in which a part of the available thermodynamic information is stored, or utilized to do work, or perhaps converted into other forms of information.

The oxidation of glucose illustrates the importance of enzymes and specific coupling mechanisms in biology. A lump of glucose can sit for years on a laboratory table, fully exposed to the air. Nothing will happen. Even though the oxidation of glucose is a spontaneous process — even though the change in Gibbs free energy produced by the reaction would be negative — even though the state of the universe after the reaction would be much more probable than the initial state, the reaction does not take place, or at least we would have to wait an enormously long time to see the glucose oxidized, because the reaction pathway is blocked by potential barriers.

Now suppose that the lump of glucose is instead eaten by a girl working in the laboratory. (She likes sweet things, and can't resist eating a lump

---

[3] The superscript ° means "under standard conditions", while kJ is an abbreviation for joule$\times 10^3$.

of sugar when she sees one.) In her body, the glucose will be oxidized almost immediately, because enzymes will lower the potential barriers along the reaction path. However, only part of the available free energy, with its content of thermodynamic information, will be degraded into heat. A large part will be coupled to the synthesis of ATP in the girl's mitochondria. The high-energy phosphate bonds of the ATP molecules will carry the available thermodynamic information further. In the end, a large part of the free energy made available by the glucose oxidation will be used to drive molecular machinery and to build up the statistically unlikely (information-containing) structures of the girl's body.

What is life?

What is Life? That was the title of a small book published by the physicist Erwin Schrödinger in 1944. Schrödinger (1887–1961) was born and educated in Austria. In 1926 he shared the Nobel Prize in Physics[4] for his contributions to quantum theory (wave mechanics). Schrödinger's famous wave equation is as fundamental to modern physics as Newton's equations of motion are to classical physics.

When the Nazis entered Austria in 1938, Schrödinger opposed them, at the risk of his life. To escape arrest, he crossed the Alps on foot, arriving in Italy with no possessions except his knapsack and the clothes which he was wearing. He traveled to England; and in 1940 he obtained a position in Ireland as Senior Professor at the Dublin Institute for Advanced Studies. There he gave a series of public lectures upon which his small book is based.

In his book, *What is Life?*, Schrödinger developed the idea that a gene is a very large information-containing molecule which might be compared to an aperiodic crystal. He also examined in detail the hypothesis (due to Max Delbrück) that X-ray induced mutations of the type studied by Hermann Muller can be thought of as photo-induced transitions from one isomeric conformation of the genetic molecule to another. Schrödinger's book has great historic importance, because Francis Crick (whose education was in physics) was one of the many people who became interested in biology as a result of reading it. Besides discussing what a gene might be in a way which excited the curiosity and enthusiasm of Crick, Schrödinger devoted a chapter to the relationship between entropy and life.

"What is that precious something contained in our food which keeps us from death? That is easily answered," Schrödinger wrote, "Every process,

---

[4] With P.A.M. Dirac.

event, happening — call it what you will; in a word, everything that is going on in Nature means an increase of the entropy of the part of the world where it is going on. Thus a living organism continually increases its entropy — or, as you may say, produces positive entropy, which is death. It can only keep aloof from it, i.e., alive, by continually drawing from its environment negative entropy — which is something very positive as we shall immediately see. What an organism feeds upon is negative entropy. Or, to put it less paradoxically, the essential thing in metabolism is that the organism succeeds in freeing itself from all the entropy it cannot help producing while alive..."[5]

"Entropy, taken with a negative sign, is itself a measure of order. Thus the device by which an organism maintains itself stationary at a fairly high level of orderliness (= fairly low level of entropy) really consists in continually sucking orderliness from its environment. This conclusion is less paradoxical than it appears at first sight. Rather it could be blamed for triviality. Indeed, in the case of higher animals we know the kind of orderliness they feed upon well enough, viz. the extremely well-ordered state of matter state in more or less complicated organic compounds which serve them as foodstuffs. After utilizing it, they return it in a very much degraded form — not entirely degraded, however, for plants can still make use of it. (These, of course, have their most powerful source of 'negative entropy' in the sunlight.)" At the end of the chapter, Schrödinger added a note in which he said that if he had been writing for physicists, he would have made use of the concept of free energy; but he judged that this concept might be difficult or confusing for a general audience.

In the paragraphs which we have quoted, Schrödinger focused on exactly the aspect of life which is the main theme of the present book: All living organisms draw a supply of thermodynamic information from their environment, and they use it to "keep aloof" from the disorder which constantly threatens them. In the case of animals, the information-containing free energy comes in the form of food. In the case of green plants, it comes primarily from sunlight. The thermodynamic information thus gained by living organisms is used by them to create configurations of matter which are so complex and orderly that the chance that they could have arisen in a random way is infinitesimally small.

John von Neumann invented a thought experiment which illustrates

---

[5] The Hungarian-American biochemist Albert Szent-Györgyi, who won a Nobel prize for isolating vitamin C, and who was a pioneer of bioenergetics, expressed the same idea in the following words: "We need energy to fight against entropy".

the role which free energy plays in creating statistically unlikely configurations of matter. Von Neumann imagined a robot or automaton, made of wires, electrical motors, batteries, etc., constructed in such a way that when floating on a lake stocked with its component parts, it will reproduce itself.[6] The important point about von Neumann's automaton is that it requires a source of free energy (i.e., a source of energy from which work can be obtained) in order to function. We can imagine that the free energy comes from electric batteries which the automaton finds in its environment. (These are analogous to the food eaten by animals.) Alternatively we can imagine that the automaton is equipped with photocells, so that it can use sunlight as a source of free energy, but it is impossible to imagine the automaton reproducing itself without some energy source from which work can be obtained to drive its reproductive machinery. If it could be constructed, would von Neumann's automaton be alive? Few people would say yes. But if such a self-reproducing automaton could be constructed, it would have some of the properties which we associate with living organisms.

The autocatalysts which are believed to have participated in molecular evolution had some of the properties of life. They used "food" (i.e., energy-rich molecules in their environments) to reproduce themselves, and they evolved, following the principle of natural selection. The autocatalysts were certainly precursors of life, approaching the borderline between non-life and life.

Is a virus alive? We know, for example, that the tobacco mosaic virus can be taken to pieces. The proteins and RNA of which it is composed can be separated, purified, and stored in bottles on a laboratory shelf. At a much later date, the bottles containing the separate components of the virus can be taken down from the shelf and incubated together, with the result that the components assemble themselves in the correct way, guided by steric and electrostatic complementarity. New virus particles are formed by this process of autoassembly, and when placed on a tobacco leaf, the new particles are capable of reproducing themselves. In principle, the stage where the virus proteins and RNA are purified and placed in bottles could be taken one step further: The amino acid sequences of the proteins and the base sequence of the RNA could be determined and written down.

Later, using this information, the parts of the virus could be synthesized from amino acids and nucleotides. Would we then be creating life? Another question also presents itself: At a certain stage in the process just

---

[6] In Chapter 8 we will return to von Neumann's self-replicating automaton and describe it in more detail.

described, the virus seems to exist only in the form of information — the base sequence of the RNA and the amino acid sequence of the proteins. Can this information be thought of as the idea of the virus in the Platonic sense? (Pythagoras would have called it the "soul" of the virus.) Is a computer virus alive? Certainly it is not so much alive as a tobacco mosaic virus. But a computer virus can use thermodynamic information (supplied by an electric current) to reproduce itself, and it has a complicated structure, containing much cybernetic information.

Under certain circumstances, many bacteria form spores, which do not metabolize, and which are able to exist without nourishment for very long periods — in fact for millions of years. When placed in a medium containing nutrients, the spores can grow into actively reproducing bacteria. There are examples of bacterial spores existing in a dormant state for many millions of years, after which they have been revived into living bacteria. Is a dormant bacterial spore alive?

Clearly there are many borderline cases between non-life and life; and Aristotle seems to have been right when he said, "Nature proceeds little by little from lifeless things to animal life, so that it is impossible to determine either the exact line of demarcation, or on which side of the line an intermediate form should lie". However, one theme seems to characterize life: It is able to convert the thermodynamic information contained in food or in sunlight into complex and statistically unlikely configurations of matter. A flood of information-containing free energy reaches the earth's biosphere in the form of sunlight. Passing through the metabolic pathways of living organisms, this information keeps the organisms far away from thermodynamic equilibrium ("which is death"). As the thermodynamic information flows through the biosphere, much of it is degraded into heat, but part is converted into cybernetic information and preserved in the intricate structures which are characteristic of life. The principle of natural selection ensures that as this happens, the configurations of matter in living organisms constantly increase in complexity, refinement and statistical improbability. This is the process which we call evolution, or in the case of human society, progress.

Suggestions for further reading

(1) S.G. Brush, *Ludwig Boltzmann and the foundations of science, in Ludwig Boltzmann Principien der Naturfilosofi*, M.I. Fasol-Boltzmann, editor, Springer, Berlin, (1990), pp. 43-64.

(2) J.C. Maxwell, *Theory of heat*, Longmans, Green and Co., London, (1902).

(3) R. A. Fisher, *On the mathematical foundations of theoretical statistics*, Phil. Trans. Roy. Soc. **222A**, 309-368 (1922).

(4) R.A. Fisher, *The Genetical Theory of Natural Selection*, Oxford University Press, (1940).

(5) R.A. Fisher, *Probability likelihood and the quantity of information in the logic of uncertain inference*, Proc. Roy. Soc. **A, 146**, 1-8 (1934)

(6) J. Neyman, *R.A. Fisher (1890-1962): An appreciation*, Science, **156**, 1456-1462 (1967).

(7) P.M. Cardozo Dias, *Clausius and Maxwell: The statistics of molecular collisions (1857-1862)*, Annals of Science, **51**, 249-261 (1994).

(8) L. Szilard, *Uber die entropieverminderung in einem thermodynamischen system bei eingriffen intelligenter wesen*, Z. Phys. **53**, 840-856 (1929).

(9) L. Szilard, *On the decrease of entropy in a thermodynamic system by the intervention of intelligent beings*, Behavioral Science **9**, 301-310 (1964).

(10) J.M. Jauch and J.G. Baron, *Entropy, information and Szilard's paradox*, Helvetica Phys. Acta, **47**, 238-247 (1974).

(11) H.S. Leff and F. Rex, editors, *Maxwell's Demon: Entropy, Information, Computing*, IOP Publishing, (1990).

(12) C.E. Shannon, *A Mathematical Theory of Communication*, Bell System Technical Journal, **27**, 379-423, 623-656, (Oct. 1948).

(13) C.E. Shannon, *Communication in the presence of noise*, Proc IRE, **37**, 10-21 (1949).

(14) C.E. Shannon and W. Weaver,*A Mathematical Theory of Communication*, University of Illinois Press, Urbana, (1949).

(15) C.E. Shannon, *Prediction and entropy in printed English*, Bell System Technical Journal, **30**, 50-64 (1951).

(16) C.E. Shannon and J. McCarthy, editors, *Automata Studies*, Princeton University Press, (1956).

(17) C.E. Shannon, *Von Neumann's contributions to automata theory*, Bull. Am. Math. Soc, **64**, 123-129 (1958).

(18) N.J.A. Sloane and C.E. Wyner, editors, *Claude Elwood Shannon: Collected Papers*, IEEE Press, New York, (1993).

(19) H. Quastler, editor, *Essays on the Use of Information Theory in Biology*, University of Illinois Press, Urbana, (1953).

(20) R.C. Raymond, *Communication, entropy and life*, American Scientist,

38, 273-278 (1950).

(21) J. Rothstein, *Information, thermodynamics and life*, Phys. Rev. **86**, 620 (1952).

(22) J. Rothstein, *Organization and entropy*, J. Appl. Phys. **23**, 1281-1282 (1952).

(23) J.R. Pierce, *An Introduction to Information Theory: Symbols, Signals and Noise*, second edition, Dover Publications, New York, (1980).

(24) L. Brillouin, *Life, thermodynamics, and cybernetics*, American Scientist, **37**, 554-568 (1949).

(25) L. Brillouin, *The negentropy principle of information*, J. Appl. Phys., **24**, 1152-1163 (1953).

(26) L. Brillouin, *Entropy and the growth of an organism*, Ann. N.Y. Acad. Sci., **63**, 454-455 (1955).

(27) L. Brillouin, *Thermodynamics, statistics, and information*, Am. J. Phys., **29**, 318-328 (1961).

(28) L. von Bertalanffy, *The theory of open systems in physics and biology*, Science, **111**, 23-29 (1950).

(29) L. von Bertalanffy, *Problems of Life*, Wiley, New York, (1952).

(30) D.A. Bell, *Physical entropy and information*, J. Appl. Phys., **23**, 372-373 (1952).

(31) F. Bonsack, *Information, Thermodynamique, Vie et Pensée*, Gauthier-Villars, Paris, (1961).

(32) K.R. Popper, *Time's arrow and feeding on negentropy*, Nature, **213**, 320 (1967).

(33) K.R. Popper, *Structural information and the arrow of time*, Nature, **214**, 322 (1967).

(34) M. Tribus and C.E. McIrvine, *Energy and Information*, Sci. Am. **225** (**3**), 179-188 (1971).

(35) F. Machlup and U. Mansfield, editors, *The Study of Information*, Wiley, New York, (1983).

(36) O. Costa de Beauregard and M. Tribus, *Information theory and thermodynamics*, Helvetica Phys. Acta, **47**, 238-247 (1974).

(37) P.W. Atkins, *The Second Law*, W.H. Freeman, N.Y., (1984).

(38) J.P. Ryan, *Aspects of the Clausius-Shannon identity: emphasis on the components of transitive information in linear, branched and composite systems*, Bull. of Math. Biol. **37**, 223-254 (1975).

(39) J.P. Ryan, *Information, entropy and various systems*, J. Theor. Biol., **36**, 139-146 (1972).

(40) R.W. Kayes, *Making light work of logic*, Nature, **340**, 19 (1970).

(41) C.H. Bennett. *The thermodynamics of computation — a review*, Int. J. Theor. Phys., **21**, 905-940 (1982).

(42) C.H. Bennett, *Demons, engines and the second law*, Sci. Am. **257** (**5**), 108-116 (1987).

(43) E.J. Chaisson, *Cosmic Evolution: The Rise of Complexity in Nature*, Harvard University Press, (2001).

(44) G.J. Eriksen and C.R. Smith, *Maximum-Entropy and Bayesian Methods in Science and Engineering*, Kluwer Academic Publishers, Dordrecht, Netherlands, (1998).

(45) C.W.F. McClare, *Chemical machines, Maxwell's demon and living organisms*, J. Theor. Biol. **30**, 1-34 (1971).

(46) G. Battail, *Does information theory explain biological evolution?*, Europhysics Letters, **40**, 343-348, (1997).

(47) T.D. Schneider, *Theory of molecular machines. I. Channel capacity of molecular machines*, J. Theor. Biol. **148**, 83-123 (1991).

(48) E.T. Jaynes, *Information Theory and Statistical Mechanics*, Phys. Rev. **106**, 620 (1957) and **108**, 171-190 (1957).

(49) R.D. Levine and M. Tribus, editors, *The Maximum Entropy Formalism*, MIT Press, Cambridge MA, (1979).

(50) B.T. Feld and G.W. Szilard (editors), *Collected Works of Leo Szilard; Scientific Papers*, The MIT Press, London and Cambridge England, (1972).

(51) A. Katz, *Principles of Statistical Mechanics — The Information Theory Approach*, Freeman, San Francisco, (1967).

(52) R. Baierlein, *Atoms and Information Theory: An Introduction to Statistical Mechanics*, Freeman, San Francisco, (1971).

(53) A. Hobson, *Concepts in Statistical Mechanics*, Gordon & Breach, New York, (1972).

(54) E. Schrödinger, *What is Life?*, Cambridge University Press, (1944).

(55) I. Prigogine, *Etude Thermodynamique des phenomènes reversible*, Dunod, Paris, (1947).

(56) I. Prigogine, *From Being to Becoming: Time and Complexity in the Physical Sciences*, W.H. Freeman, San Francisco, (1980).

(57) I. Prigogine and K. Stegers, *Order Out of Chaos: Man's New Dialogue With Nature*, Bantam, New York, (1984).

(57) L.L. Gatlin, *The information content of DNA*, J. Theor. Biol. 10, 281-300 (1966), and 18, 181-194 (1968).

(58) J. von Neumann, *Theory of self-replicating automata*, University of Illinois Press, Urbana, (1966).

(59) J. von Neumann, *Probabilistic logics and the synthesis of reliable organisms from unreliable components*, in *Collected works* (A. Taub editor), **vol. 5**, pp. 341-347, MacMillan, New York, (1963).

(60) P. Morison, *A thermodynamic characterization of self-reproduction*, Rev. Mod. Phys. **36**, 517-564 (1964).

(61) C.I.J.M. Stuart, *Physical models of biological information and adaption*, J. Theor. Biol., **113**, 441-454 (1985).

(62) C.I.J.M. Stuart, *Bio-informational equivalence*, J. Theor. Biol., 113, 611-636 (1985).

(63) R.E. Ulanowicz and B.M. Hannon, *Life and the production of entropy*, Proc. Roy. Soc. Lond., **ser. B, 32**, 181-192 (1987).

(64) D.R. Brooks and E.O. Wilson, *Evolution as Entropy: Toward a Unified Theory of Biology*, University of Chicago Press, (1986).

(65) H.T. Odum, *Self-organization, transformity and information*, Science, 242, 1132-1139 (1988).

(66) B. Weber, D. Depew and J. Smith, editors, *Entropy, Information, and Evolution: New Perspectives on Physical and Biological Evolution*, MIT Press, Cambridge MA, (1988).

(67) R.U. Ayres, *Information, Entropy and Progress: A New Evolutionary Paradigm*, AIP Press, New York, (1994).

(68) R.H. Vreeland, W.T. Rosenzweig and D.W. Powers, *Isolation of a 250 million-year-old halotolerant bacterium from a primary salt crystal*, Nature, **407**, 897-900 (19 October 2000).

(69) Albert Szent-Györgyi, *Bioenergetics*, Academic Press, New York, (1957).

(70) A.L. Leheninger, *Bioenergetics*, W.A. Benjamin, New York, (1965).

(71) J. Avery (editor), *Membrane Structure and Mechanisms of Biological Energy Transduction*, Plenum Press, London, (1974).

(72) T.J. Hill, *Free Energy Transduction in Biology*, Academic Press, (1977).

(73) J. Avery, *A model for the primary process in photosynthesis*, Int. J. Quant. Chem., **26**, 917 (1984).

(74) D.G. Nicholls and S.J. Furgason, *Bioenergetics 2*, Academic Press, (1992).

Chapter 5

INFORMATION FLOW IN BIOLOGY

Cybernetic (or semiotic) information — codes and languages

As we have seen in Chapter 4, the evolutionary process is driven by an enormous flow of thermodynamic information passing through the earth's biosphere. Most of it is lost — degraded into heat — but a tiny fraction of the flow is stabilized and preserved as cybernetic information[1]. This second form of information, which is associated with the sending and receiving of signals, with communication, with codes or languages, and with biological or cultural complexity, will be the theme of the remaining chapters of this book[2]. In addition to the flow of Gibbs free energy (i.e., thermodynamic information) through living organisms and ecosystems, there is also a flow of cybernetic (or semiotic) information.

The language of molecular complementarity

In living (and even non-living) systems, signals can be written and read at the molecular level. The language of molecular signals is a language of complementarity. The first scientist to call attention to complementarity and pattern recognition at the molecular level was Paul Ehrlich, who was born in 1854 in Upper Silesia (now a part of Poland). Ehrlich was not an especially good student, but his originality attracted the attention of his teacher, Professor Waldeyer, under whom he studied chemistry at the University of Strasbourg. Waldeyer encouraged him to do independent ex-

---

[1] The polymerase chain reaction discussed in Chapter 3 can be thought of as a process in which thermodynamic information is converted into cybernetic information. A second example is Spiegelman's experiment, discussed at the end of Chapter 8.

[2] Sometimes information associated with signs is alternatively called "semiotic information" as is discussed in Appendix 2.

periments with the newly-discovered aniline dyes; and on his own initiative, Ehrlich began to use these dyes to stain bacteria. He was still staining cells with aniline dyes a few years later (by this time he had become a medical student at the University of Breslau) when the great bacteriologist Robert Koch visited the laboratory. "This is young Ehrlich, who is very good at staining, but will never pass his examinations", Koch was told. Nevertheless, Ehrlich did pass his examinations, and he went on to become a doctor of medicine at the University of Leipzig at the age of 24. His doctoral thesis dealt with the specificity of the aniline dyes: Each dye stained a special class of cell and left all other cells unstained.

Paul Ehrlich had discovered what might be called "the language of molecular complementarity": He had noticed that each of his aniline dyes stained only a particular type of tissue or a particular species of bacteria. For example, when he injected one of his blue dyes into the ear of a rabbit, he found to his astonishment that the dye molecules attached themselves selectively to the nerve endings. Similarly, each of the three types of phagocytes could be stained with its own particular dye, which left the other two kinds unstained[3].

Ehrlich believed that this specificity came about because the side chains on his dye molecules contained groupings of atoms which were complementary to groups of atoms on the surfaces of the cells or bacteria which they selectively stained. In other words, he believed that biological specificity results from a sort of lock and key mechanism: He visualized a dye molecule as moving about in solution until it finds a binding site which exactly fits the pattern of atoms in one of its side chains. Modern research has completely confirmed this picture, with the added insight that we now know that the complementarity of the "lock" and "key" is electrostatic as well as spatial.

Two molecules in a biological system may fit together because the contours of one are complementary to the contours of the other. This is how Paul Ehrlich visualized the fit — a spatial (steric) complementarity, like that of a lock and key. However, we now know that for maximum affinity, the patterns of excess charges on the surfaces of the two molecules must

---

[3] The specificity which Ehrlich observed in his staining studies made him hope that it might be possible to find chemicals which would attach themselves selectively to pathogenic bacteria in the blood stream and kill the bacteria without harming normal body cells. He later discovered safe cures for both sleeping sickness and syphilis, thus becoming the father of chemotherapy in medicine. He had already received the Nobel Prize for his studies of the mechanism of immunity, but after his discovery of a cure for syphilis, a street in Frankfurt was named after him!

also be complementary. Regions of positive excess charge on the surface of one molecule must fit closely with regions of negative excess charge on the other if the two are to bind maximally. Thus the language of molecules is not only a language of contours, but also a language of charge distributions.

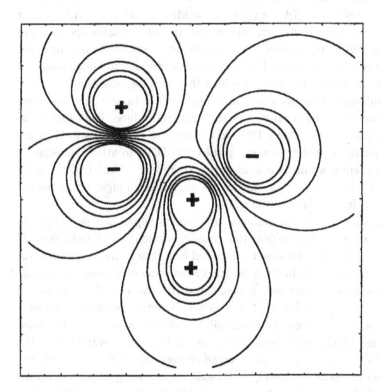

Fig. 5.1 This figure shows the excess charges and the resulting electrostatic potential on a molecule of formic acid, HCOOH. The two oxygens in the carboxyl group are negatively charged, while the carbon and the two hydrogens have positive excess charges. Molecular recognition involves not only steric complementarity, but also complementarity of charge patterns.

The flow of information between and within cells

Information is transferred between cells in several ways. Among bacteria, in addition to the chronologically vertical transfer of genetic information directly from a single parent to its two daughter cells on cell division, there are mechanisms for the sharing of genetic information in a chronologically horizontal way, between cells of the same generation. These horizontal genetic information transfers can be thought of as being analogous to sex, as will be seen more clearly from some examples.

In the most primitive mechanism of horizontal information transfer, a bacterium releases DNA into its surroundings, and the DNA is later absorbed by another bacterium, not necessarily of the same species. For example, a loop or plasmid of DNA conferring resistance to an antibiotic (an "R-factor") can be released by a resistant bacterium and later absorbed by a bacterium of another species, which then becomes resistant[4].

A second mechanism for horizontal information transfer involves infection of a bacterium by a virus. As the virus reproduces itself inside the bacterium, some of the host's DNA can chance to be incorporated in the new virus particles, which then carry the extra DNA to other bacteria.

Finally, there is a third mechanism (discovered by Joshua Lederberg) in which two bacteria come together and construct a conjugal bridge across which genetic information can flow.

Almost all multicellular animals and plants reproduce sexually. In the case of sexual reproduction the genetic information of both parents is thrown into a lottery by means of special cells, the gametes. Gametes of each parent contain only half the genetic information of the parent, and the exact composition of that half is determined by chance. Thus, when the gametes from two sexes fuse to form a new individual, the chances for variability are extremely large. This variability is highly valuable to multicellular organisms which reproduce sexually, not only because variability is the raw material of evolutionary adaption to changes in the environment, but also because the great variability of sexually-reproducing organisms makes them less likely to succumb to parasites. Infecting bacteria might otherwise deceive the immune systems of their hosts by developing cell-surface antigens which resemble those of the host, but when they infect sexually-reproducing organisms where each individual is unique, this is much less likely.

---

[4] The fact that this can happen is a strong reason for using antibiotics with great caution in agriculture. Resistance to antibiotics can be transferred from the bacteria commonly found in farm animals to bacteria which are dangerous for humans. Microbiologists have repeatedly warned farmers, drug companies and politicians of this danger, but the warnings have usually been ignored. Unfortunately there are now several instances of antibiotic-resistant human pathogens that have been produced by indiscriminate use of antibiotics in agriculture.

Within the cells of all organisms living today, there is a flow of information from polynucleotides (DNA and RNA) to proteins. As messenger RNA passes through a ribosome, like punched tape passing through a computer tapereader, the sequence of nucleotides in the mRNA is translated into the sequence of nucleic acids in the growing protein. The molecular mechanism of the reading and writing in this process involves not only spatial complementarity, but also complementarity of charge distributions.

As a protein grows, one amino acid at a time, it begins to fold. The way in which it folds (the "tertiary conformation") is determined both by spatial complementarity and by complementarity of charge distributions: Those amino acids which have highly polar groups, i.e., where several atoms have large positive or negative excess charges — "hydrophilic" amino acids — tend to be placed on the outside of the growing protein, while amino acids lacking large excess charges — "hydrophobic" amino acids — tend to be on the inside, away from water. Hydrophilic amino acids form hydrogen bonds with water molecules. Whenever there is a large negative charge on an atom of an amino acid, it attracts a positively-charged hydrogen from water, while positively-charged hydrogens on nucleic acids are attracted to negatively charged oxygens of water. Meanwhile, in the interior of the growing protein, non-polar amino acids are attracted to each other by so-called van der Waals forces, which do not require large excess charges, but only close proximity.

When a protein is complete, it is ready to participate in the activities of the cell, perhaps as a structural element or perhaps as an enzyme. Enzymes catalyze the processes by which carbohydrates, and other molecules used by the cell, are synthesized. Often an enzyme has an "active site", where such a process takes place. Not only the spatial conformation of the active site but also its pattern of excess charges must be right if the catalysis is to be effective. An enzyme sometimes acts by binding two smaller molecules to its active site in a proper orientation to allow a reaction between them to take place. In other cases, substrate molecules are stressed and distorted by electrostatic forces as they are pulled into the active site, and the activation energy for a reaction is lowered.

Thus, information is transferred first from DNA and RNA to proteins, and then from proteins to (for example) carbohydrates. Sometimes the carbohydrates then become part of surface of a cell. The information which these surface carbohydrates ("cell surface antigens") contain may be transmitted to other cells. In this entire information transfer process, the "reading" and "writing" depend on steric complementarity and on complementarity of molecular charge distributions.

Not only do cells communicate by touching each other and recognizing each other's cell surface antigens — they also communicate by secreting and absorbing transmitter molecules. For example, the group behavior of slime mold cells is coordinated by the cyclic adenosine monophosphate molecules, which the cells secrete when distressed.

Within most multicellular organisms, cooperative behavior of cells is coordinated by molecules such as hormones — chemical messengers. These are recognized by "receptors", the mechanism of recognition once again depending on complementarity of charge distributions and shape. Receptors on the surfaces of cells are often membrane-bound proteins which reach from the exterior of the membrane to the interior. When an external transmitter molecule is bound to a receptor site on the outside part of the protein, it causes a conformational change which releases a bound molecule of a different type from a site on the inside part of the protein, thus carrying the signal to the cell's interior. In other cases the messenger molecule passes through the cell membrane.

In this way, the individual cell in a society of cells (a multicellular organism) is told when to divide and when to stop dividing, and what its special role will be in the economy of the cell society (differentiation). For example, in humans, follicle-stimulating hormone, lutenizing hormone, prolactin, estrogen and progesterone are among the chemical messengers which cause the cell differentiation needed to create the secondary sexual characteristics of females.

Another role of chemical messengers in multicellular organisms is to maintain a reasonably constant internal environment in spite of drastic changes in the external environment of individual cells or of the organism as a whole (homeostasis). An example of such a homeostatic chemical messenger is the hormone insulin, which is found in humans and other mammals. The rate of its release by secretory cells in the pancreas is increased by high concentrations of glucose in the blood. Insulin carries the news of high glucose levels to target cells in the liver, where the glucose is converted to glycogen, and to other target cells in the muscles, where the glucose is burned.

Nervous systems

Hormones require a considerable amount of time to diffuse from the cells where they originate to their target cells; but animals often need to act very quickly, in fractions of seconds, to avoid danger or to obtain food. Because of the need for quick responses, a second system of communication has evolved — the system of neurons.

Neurons have a cell bodies, nuclei, mitochondria and other usual features of eukaryotic cells, but in addition they possess extremely long and thin tubelike extensions called axons and dendrites. The axons function as informational output channels, while the dendrites are inputs. These very long extensions of neurons connect them with other neurons which can be at distant sites, to which they are able to transmit electrical signals. The complex network of neurons within a multicellular organism, its nervous system, is divided into three parts. A sensory or input part brings in signals from the organism's interior or from its external environment. An effector or output part produces a response to the input signal, for example by initiating muscular contraction. Between the sensory and effector parts of the nervous system is a message-processing (internuncial) part, whose complexity is not great in the jellyfish or the leech. However, the complexity of the internuncial part of the nervous system increases dramatically as one goes upward in the evolutionary order of animals, and in humans it is truly astonishing.

The small button-like connections between neurons are called synapses. When an electrical signal propagating along an axon reaches a synapse, it releases a chemical transmitter substance into the tiny volume between the synapse and the next neuron (the post-synaptic cleft). Depending on the nature of the synapse, this chemical messenger may either cause the next neuron to "fire" (i.e., to produce an electrical pulse along its axon) or it may inhibit the firing of the neuron. Furthermore, the question of whether a neuron will or will not fire depends on the past history of its synapses. Because of this feature, the internuncial part of an animal's nervous system is able to learn. There many kinds of synapses and many kinds of neurotransmitters, and the response of synapses is sensitive to the concentration of various molecules in the blood, a fact which helps to give the nervous systems of higher animals extraordinary subtlety and complexity.

The first known neurotransmitter molecule, acetylcholine, was discovered jointly by Sir Henry Dale in England and by Otto Loewi in Germany. In 1921, Loewi was able to show that nerve endings transmit information to muscles by means of this substance. The idea for the critical experiment occurred to him in a dream at 3 am. Otto Loewi woke up and wrote down the idea; but in the morning he could not read what he had written.

Luckily he had the same dream the following night. This time he took no chances. He got up, drank some coffee, and spent the whole night working in his laboratory. By morning he had shown that nerve cells separated from the muscle of a frog's heart secrete a chemical substance when stimulated, and that this substance is able to cause contractions of the heart of another frog. Sir Henry Dale later showed that Otto Loewi's transmitter molecule was identical to acetylcholine, which Dale had isolated from the ergot fungus in 1910. The two men shared a Nobel Prize in 1936. Since that time, a large variety of neurotransmitter molecules have been isolated. Among the excitatory neurotransmitters (in addition to acetylcholine) are noradrenalin, norepinephrine, serotonin, dopamine, and glutamate, while gamma-amino-butyric acid is an example of an inhibitory neurotransmitter.

The mechanism by which electrical impulses propagate along nerve axons was clarified by the English physiologists Alan Lloyd Hodgkin and Andrew Fielding Huxley (a grandson of Darwin's defender, Thomas Henry Huxley). In 1952, working with the giant axon of the squid (which can be as large as a millimeter in diameter), they demonstrated that the electrical impulse propagating along a nerve is in no way similar to an electrical current in a conducting wire, but is more closely analogous to a row of dominoes knocking each other down. The nerve fiber, they showed, is like a long thin tube, within which there is a fluid containing K^+ and Na^+ ions, as well as anions. Inside a resting nerve, the concentration of K^+ is higher than in the normal body fluids outside, and the concentration of Na^+ is lower. These abnormal concentrations are maintained by an "ion pump", which uses the Gibbs free energy of adenosine triphosphate (ATP) to bring potassium ions into the nerve and to expel sodium ions.

The membrane surrounding the neural axon is more permeable to potassium ions than to sodium, and the positively charged potassium ions tend to leak out of the resting nerve, producing a small difference in potential between the inside and outside. This "resting potential" helps to hold the molecules of the membrane in an orderly layer, so that the membrane's permeability to ions is low.

Hodgkin and Huxley showed that when a neuron fires, the whole situation changes dramatically. Triggered by the effects of excitatory neurotransmitter molecules, sodium ions begin to flow into the axon, destroying the electrical potential which maintained order in the membrane. A wave of depolarization passes along the axon. Like a row of dominoes falling, the disturbance propagates from one section to the next: Sodium ions flow in,

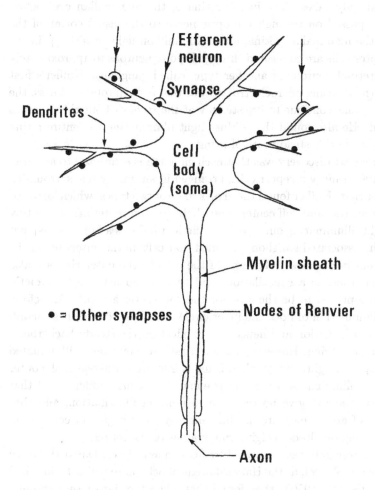

Fig. 5.2   A schematic diagram of a neuron.

the order-maintaining electrical potential disappears, the next small section of the nerve membrane becomes permeable, and so on. Thus, Hodgkin and Huxley showed that when a neuron fires, a quick pulse-like electrical and chemical disturbance is transmitted along the axon.

In 1953, Stephen W. Kuffler, working at Johns Hopkins University, made a series of discoveries which yielded much insight into the mechanisms by which the internuncial part of mammalian nervous systems processes information. Kuffler's studies showed that some degree of abstraction of patterns already takes place in the retina of the mammalian eye, before signals are passed on through the optic nerve to the visual cortex of the brain. In the mammalian retina, about 100 million light-sensitive primary light-receptor cells are connected through bipolar neurons to approximately a million retinal neurons of another type, called ganglions. Kuffler's first discovery (made using microelectrodes) was that even in total darkness, the retinal ganglions continue to fire steadily at the rate of about thirty pulses per second. He also found that diffuse light illuminating the entire retina does not change this steady rate of firing.

Kuffler's next discovery was that each ganglion is connected to an array of about 100 primary receptor cells, arranged in an inner circle surrounded by an outer ring. Kuffler found the arrays to be of two types, which he called "on center arrays" and "off center arrays". In the "on center arrays", a tiny spot of light, illuminating only the inner circle, produces a burst of frequent firing of the associated ganglion, provided that cells in the outer ring of the array remain in darkness. However, if the cells in the outer ring are also illuminated, there is a cancellation, and there is no net effect. Exactly the opposite proved to be the case for the "off center arrays". As before, uniform illumination of both the inner circle and outer ring of these arrays produces a cancellation and hence no net effect on the steady background rate of ganglion firing. However, if the central circle by itself is illuminated by a tiny spot of light, the ganglion firing is inhibited, whereas if the outer ring alone is illuminated, the firing is enhanced. Thus Kuffler found that both types of arrays give no response to uniform illumination, and that both types of arrays measure, in different ways, the degree of contrast in the light falling on closely neighboring regions of the retina.

Kuffler's research was continued by his two associates, David H. Hubel and Torsten N. Wessel, at the Harvard Medical School, to which Kuffler had moved. In the late 1950's, they found that when the signals sent through the optic nerves reach the visual cortex of the brain, a further abstraction of patterns takes place through the arrangement of connections between two successive layers of neurons. Hubel and Wessel called the cells in these two pattern-abstracting layers "simple" and "complex". The retinal ganglions were found to be connected to the "simple" neurons in such a way that a "simple" cell responds to a line of contrasting illumination of the retina. For such a cell to respond, the line has to be at a particular position and has to have a particular direction. However, the "complex" cells in the next layer were found to be connected to the "simple" cells in such a way that they respond to a line in a particular direction, even when it is displaced parallel to itself[5].

In analyzing their results, Kuffler, Hubel and Wessel concluded that pattern abstraction in the mammalian retina and visual cortex takes place through the selective destruction of information. This conclusion agrees with what we know in general about abstractions: They are always simpler than the thing which they represent.

Animal languages

Communication between two or more multicellular organism often takes place through the medium of signal molecules, which are recognized by receptors. For example, the perfume of flowers is recognized by insects (and by us). Insect pheromones are among the most powerful signal molecules.

The language of ants depends predominantly on chemical signals. In most mammals too, the sense of smell plays a large role in mating, maternal behavior, and group organization[6]. Anyone who has owned a pet cat or dog knows what an important role the sense of smell plays in their social lives.

Pheromones are defined as chemical compounds that are exchanged as signals between members of the same species, and very many of these substances have now been isolated and studied. Pheromones often play a role in reproduction. For example, females of the silkworm moth species Bombyx mori emit an alcohol, *trans*-10-*cis*-12-hexadecadienol, from a gland in tip of their their abdomens. The simplified name of this alcohol is "bombykol", after the name of the moth. The male moth is equipped with feathery antennae, the hairs of which are sensitive to the pheromone — so sensitive in fact that a receptor on one of the hairs is able to register the presence of a single bombykol molecule! Aroused by even a very modest concentration of bombykol, the male finds himself compelled by the inherited programs of his brain to follow the path of increasing concentration until he finds the female and mates with her.

---

[5] Interestingly, at about the same time, the English physiologist J.Z. Young came to closely analogous conclusions regarding the mechanism of pattern abstraction in the visual cortex of the octopus brain. However, the similarity between the image-forming eye of the octopus and the image-forming vertebrate eye and the rough similarity between the mechanisms for pattern abstraction in the two cases must both be regarded as instances of convergent evolution, since the mollusc eye and the vertebrate eye have evolved independently.

[6] Puppies up to the age of 7 weeks or so have a distinctive odor which is attractive to humans as well as to dogs.

The pheromone *trans*-9-keto-2-decanoic acid, the "queen substance", plays a somewhat more complex role in the social organization of the honeybee. This pheromone, which is emitted by the queen's mandibular glands, has several functions. Workers lick the queen's body and regurgitate the substance back and forth to each other, so that it is spread throughout the hive. When they do so, their ovaries fail to develop, and they are also restrained from raising larvae in such a way that the young bees could become queens. Thus, as long as the reigning queen is alive and producing the pheromone, she has no rivals. Another function of *trans*-9-keto-2-decanoic acid is to guide a husband to the queen on her nuptial flight and to promote the consummation of their marriage.

Worker bees cannot recognize each other as individuals, but each hive has a distinctive scent, shared by all its members. Foreign bees, with a different nest scent, are aggressively repelled. Like bees, their close relatives the ants also have a distinctive nest scent by which members of a colony recognize each other and repel foreigners.

Ants use chemical trails to guide each other to sources of food. An ant which has found an open jam jar marks the trail to it with a signalling substance, and other ants following this pheromone trail increase the intensity of the marking. However, the signal molecules continually evaporate. Eventually the trails disappear, and the ants are freed to explore other sources of food.

Bees guide each other to sources of food by another genetically programmed signaling method — the famous waggle dance, deciphered in 1945 by Karl von Frisch. When a worker bee has found a promising food source, she returns to the hive and performs a complex dance, the pattern of which indicates both the direction and distance of the food. The dancer moves repeatedly in a pattern resembling the Greek letter Θ. If the food-discoverer is able to perform her dance on a horizontal flat surface in view of the sun, the line in the center of the pattern points in the direction of the food. However, if the dance is performed in the interior of the hive on a vertical surface, gravity takes the place of the sun, and the angle between the central line and the vertical represents the angle between the food source and the sun.

The central part of the dance is, in a way, a re-enactment of the excited forager's flight to the food. As she traverses the central portion of the pattern, she buzzes her wings and waggles her abdomen rapidly, the number of waggles indicating the approximate distance to the food[7]. After this central portion of the dance, she turns alternately to the left or to the right, following one or the other of the semicircles, and repeats the performance. Studies of the accuracy with which her hive-mates follow these instructions show that the waggle dance is able to convey approximately seven bits of information — three bits concerning distance and four bits concerning direction. After making his initial discovery of the meaning of the dance, von Frisch studied the waggle dance in many species of bees. He was able to distinguish species-specific dialects, and to establish a plausible explanation for the evolution of the dance.

Like bees, most mammals have communication systems which utilize not only scent, but also other displays and signals. For example, galagos or bushbabies, small furry primates found in the rainforests of Africa, have scent glands on their faces, chests, arms, elbows, palms, and soles, and they also scent-mark their surroundings and each other with saliva and urine. In fact, galagos bathe themselves in urine, standing on one foot and using their hands and feet as cups. This scent-repertoire is used by the bushbabies to communicate reproductive and social information. However, in addition, they also communicate through a variety of calls. They croak, chirp, click, whistle and bark, and the mating call of the Greater Galago sounds exactly like the crying of a baby — whence the name.

The communication of animals (and humans) through visual displays was discussed by Charles Darwin in his book *The Expression of the Emotions in Man and Animals*. For example, he discussed the way in which the emotions of a dog are expressed as visual signs: "When a dog approaches a strange dog or man in a savage or hostile frame of mind", Darwin wrote, "he walks very stiffly; his head is slightly raised, or not much lowered; the tail is held erect and quite rigid; the hairs bristle, especially along the neck and back; the pricked ears are directed forwards, and the eyes have a fixed stare... Let it now be supposed that the dog suddenly discovers that the man he is approaching is not a stranger, but his master; and let it be observed how completely and instantaneously his whole bearing is reversed. Instead of walking upright, the body sinks downwards or even crouches, and is thrown into flexuous movements; the tail, instead of being held stiff and upright, is lowered and wagged from side to side; his hair instantly becomes smooth; his ears are depressed and drawn backwards, but not closely to the head, and his lips hang loosely. From the drawing back of the ears,

---

[7] The number of waggles is largest when the source of food is near, and for extremely nearby food, the bees use another dance, the "round dance".

the eyelids become elongated, and the eyes no longer appear round and staring".

A wide variety of animals express hostility by making themselves seem larger than they really are: Cats arch their backs, and the hairs on their necks and backs are involuntarily raised; birds ruffle their feathers and spread their wings; lizards raise their crests and lower their dewlaps; and even some species of fish show hostility by making themselves seem larger, by spreading their fins or extending their gill covers. Konrad Lorenz has noted, in his book *On Aggression*, that the "holy shiver" experienced by humans about to perform an heroic act in defense of their community is closely related to the bristling hair on the neck and back of a cat or dog when facing an enemy.

Human language has its roots in the nonverbal signs by which our evolutionary predecessors communicated, and traces of early human language can be seen in the laughter, tears, screams, groans, grins, winks, frowns, sneers, smiles, and explanatory gestures which we use even today to clarify and emphasize our words.

Suggestions for further reading

(1) M. Eigen et al., *The Origin of genetic information*, Scientific American, April, 78-94 (1981).

(2) L.E. Kay, *Cybernetics, information, life: The emergence of scriptural representations of heredity*, Configurations, **5**, 23-91 (1997).

(3) T.D. Schneider, G.D. Stormo, L. Gold and A. Ehrenfeucht, *Information content of binding sites on nucleotide sequences*, J. Mol. Biol. **88**, 415-431 (1986).

(4) J. Avery, *A model for biological specificity*, Int. J. Quant. Chem., **26**, 843 (1984).

(5) P.G. Mezey, *Shape in Chemistry: An Introduction to Molecular Shape and Topology*, VCH Publishers, New York, (1993).

(6) P.G. Mezey, *Potential Energy Hypersurfaces*, Elsevier, Amsterdam, (1987).

(7) P.G. Mezey, *Molecular Informatics and Topology in Chemistry*, in *Topology in Chemistry*, R.B. King and D.H. Rouvray, eds., Ellis Horwood, Pbl., U.K., (2002).

(8) G. Stent, *Cellular communication*, Scientific American, **227**, 43-51 (1972).

(9) A. Macieira-Coelho, editor, *Signaling Through the Cell Matrix*,

Progress in Molecular and Subcellular Biology, **25**, Springer, (2000).

(10) D.H. Hubel, *The visual cortex of the brain*, Scientific American, 209, 54, November, (1963).

(11) G. Stent, editor, *Function and Formation of Neural Systems.*

(12) J.Z. Young, *Programs of the Brain*, Oxford University Press, (1978).

(13) J.Z. Young, *Philosophy and the Brain*, Oxford University Press, (1987).

(14) K. von Frisch, *Dialects in the languages of bees*, Scientific American, August, (1962).

(15) R.A. Hinde, *Non-Verbal Communication*, Cambridge University Press, (1972).

(16) E.O. Wilson, *Animal communication*, Scientific American, **227**, 52-60 (1972).

(17) E.O. Wilson, *Sociobiology*, Harvard University Press, (1975).

(18) H.S. Terrace, L.A. Petitto, et al., *Can an ape create a sentence?*, Science, **206**, 891-902 (1979).

(19) S. Savage-Rumbaugh, R. Lewin, et al., Kanzi: *The Ape at the Brink of the Human Mind*, John Wiley and Sons, New York, (1996).

(20) R.W. Rutledge, B.L. Basore, and R.J. Mulholland, *Ecological stability: An information theory viewpoint*, J. Theor. Biol., **57**, 355-371 (1976).

(21) L. Johnson, *Thermodynamics and ecosystems*, in *The Handbook of Environmental Chemistry*, O. Hutzinger, editor, Springer Verlag, Heidelberg, (1990), pp. 2-46.

(22) C. Pahl-Wostl, *Information theoretical analysis of functional temporal and spatial organization in flow networks*, Math. Comp. Model. **16 (3)**, 35-52 (1992).

(23) C. Pahl-Wostl, *The Dynamic Nature of Ecosystems: Chaos and Order Intertwined*, Wiley, New York, (1995).

(24) E.D. Schneider and J.J. Kay, *Complexity and thermodynamics: Towards a new ecology*, Futures, **24 (6)**, 626-647 (1994).

(25) R.E. Ulanowicz,*Ecology, the Ascendent Perspective*, Colombia University Press, New York, (1997).

Chapter 6

CULTURAL EVOLUTION AND INFORMATION

The coevolution of human language, culture, and intelligence

The prehistoric genetic evolution of modern humans, as well as their more recent cultural evolution can be understood in terms of a single theme — information. The explosively rapid development of our species can be thought of as a continually accelerating accumulation of information, as this chapter will try to demonstrate.

In his *Systema Naturae*, published in 1735, Carolus Linnaeus correctly classified humans as mammals associated with the anthropoid apes. However, illustrations of possible ancestors of humans in a later book by Linnaeus, showed one with a manlike head on top of a long-haired body, and another with a tail. A century later, in 1856, light was thrown on human ancestry by the discovery of some remarkable bones in a limestone cave in the valley of Neander, near Düsseldorf — a skullcap and some associated long bones. The skullcap was clearly manlike, but the forehead was low and thick, with massive ridges over the eyes. The famous pathologist Rudolf Virchow dismissed the find as a relatively recent pathological idiot. Other authorities thought that it was "one of the Cossacks who came from Russia in 1814". Darwin knew of the "Neanderthal man", but he was too ill to travel to Germany and examine the bones. However, Thomas Huxley examined them, and in his 1873 book, Zoological Evidences of Man's Place in Nature, he wrote: "Under whatever aspect we view this cranium... we meet with apelike characteristics, stamping it as the most pithecoid (apelike) of human crania yet discovered".

"In some older strata," Huxley continued, "do the fossilized bones of an ape more anthropoid, or a man more pithecoid, than any yet known await the researches of some unborn paleontologist?" Huxley's question obsessed Eugene Dubois, a young Dutch physician, who reasoned that such a find would be most likely in Africa, the home of chimpanzees and gorillas, or in the East Indies, where orang-utans live. He was therefore happy to be appointed to a post in Sumatra in 1887. While there, Dubois heard of a site in Java where the local people had discovered many ancient fossil bones, and at this site, after much searching, he uncovered a cranium which was much too low and flat to have belonged to a modern human. On the other hand it had features which proved that it could not have belonged to an ape. Near the cranium, Dubois found a leg bone which clearly indicated upright locomotion, and which he (mistakenly) believed to belong to the same creature. In announcing his find in 1894, Dubois proposed the provocative name "Pithecanthropus erectus", i.e., "upright-walking ape-man".

Instead of being praised for this discovery, Dubois was denounced. His attackers included not only the clergy, but also many scientists (who had expected that an early ancestor of man would have an enlarged brain associated with an apelike body, rather than apelike head associated with upright locomotion). He patiently exhibited the fossil bones at scientific meetings throughout Europe, and gave full accounts of the details of the site where he had unearthed them. When the attacks nevertheless continued, Dubois became disheartened, and locked the fossils in a strongbox, out of public view, for the next 28 years. In 1923, however, he released a cast of the skull, which showed that the brain volume was about 900 cm^3 — well above the range of apes, but below the 1200–1600 cm^3 range which characterizes modern man. Thereafter he again began to exhibit the bones at scientific meetings.

The fossil bones of about 1000 hominids, intermediate between apes and humans, have now been discovered. The oldest remains have been found in Africa. Many of these were discovered by Raymond Dart and Robert Broom, who worked in South Africa, and by Louis and Mary Leaky and their son Richard, who made their discoveries at the Olduvai Gorge in Tanzania and at Lake Rudolph in Kenya. Table 6.1 shows some of the more important species and their approximate dates.

One can deduce from biochemical evidence that the most recent common ancestor of the anthropoid apes and of humans lived in Africa between 5 and 10 million years before the present. Although the community of palaeoanthropologists is by no means unanimous, there is reasonably general agreement that while A. africanus is probably an ancestor of H. habilis and of humans, the "robust" species, A. aethiopicus, A. robustus and A.

Table **6.1**: Hominid species

Genus and species	Years before present	Brain volume
Ardipithicus ramidus	5.8 to 4.4 million	
Australopithecus anamensis	4.2 to 3.9 million	
Australopithecus afarensis	3.9 to 3.0 million	375 to 550 cm^3
Australopithecus africanus	3 to 2 million	420 to 500 cm^3
Australopithecus aethiopicus	2.6 to 2.3 million	410 cm^3
Australopithecus robustus	2 to 1.5 million	530 cm^3
Australopithecus boisei	2.1 to 1.1 million	530 cm^3
Homo habilis	2.4 to 1.5 million	500 to 800 cm^3
Homo erectus	1.8 to 0.3 million	750 to 1225 cm^3
Homo sapiens (archaic)	0.5 to 0.2 million	1200 cm^3
Homo sapiens neand.	0.23 to 0.03 million	1450 cm^3
Homo sapiens sapiens	0.12 mil. to present	1350 cm^3

boisei[1] represent a sidebranch which finally died out. "Pithecanthropus erectus", found by Dubois, is now classified as a variety of Homo erectus, as is "Sinanthropus pekinensis" ("Peking man"), discovered in 1929 near Beijing, China.

Footprints 3.7 million years old showing upright locomotion have been

---

[1] A. boisei was originally called "Zinjanthropus boisei" by Mary and Louis Leakey who discovered the fossil remains at the Olduvai Gorge. Charles Boise helped to finance the Leakey's expedition.

Table 6.2: Paleolithic cultures

Name	Years before present	Characteristics
Oldowan	2.4 to 1.5 million	Africa, flaked pebble tools
Choukoutien	1.2 to 0.5 million	chopper tool culture of east Asia
Abbevillian	500,000 to 450,000	crude stone handaxes
Mousterian	70,000 to 20,000	Africa, Europe, northeast Asia produced by Neanderthal man, retouched core and flake tools, wooden spears, fire, burial of dead
Aurignacian	50,000 to 20,000	western Europe, fine stone blades, pins and awls of bone, fire, cave art
Solutrian	20,000 to 17,000	France and central Europe, long, pressure-flaked bifacial blades
Magdalenian	17,000 to 10,000	western Europe, reindeer hunting awls and needles of bone and antler

discovered near Laetoli in Tanzania. The Laetoli footprints are believed to have been made by A. afarensis, which was definitely bipedal, but upright locomotion is thought to have started much earlier. There is even indirect evidence which suggests that A. ramidus may have been bipedal. Homo habilis was discovered by Mary and Louis Leakey at the Olduvai Gorge, among beds of extremely numerous pebble tools. The Leakeys gave this name (meaning "handy man") to their discovery in order to call special attention to his use of tools. The brain of H. habilis is more human than that of A. africanus, and in particular, the bulge of Broca's area, essential for speech, can be seen on one of the skull casts. This makes it seem likely that H. habilis was capable of at least rudimentary speech.

Homo erectus was the first species of hominid to leave Africa, and his remains are found not only there, but also in Europe and Asia. "Peking man", who belonged to this species, probably used fire. The stone tools of H. erectus were more advanced than those of H. habilis; and there is no sharp line of demarcation between the most evolved examples of H. erectus and early fossils of archaic H. sapiens.

Homo sapiens neanderthalensis lived side by side with Homo sapiens sapiens (modern man) for a hundred thousand years; but in relatively recent times, only 30,000 years ago, Neanderthal man disappeared. Recently Svante Pääbo of the Max Planck Institute for Evolutionary Anthropology and his group were able sequence the Neanderthal genome. In 2010 Pääbo announced that between 1% and 4% of the genes of modern humans outside Africa are of Neanderthal origin, the implication being that some interbreeding took place between Neanderthals and modern humans.

More recently, Svante Pääbo's group was able to sequence the DNA from a tooth and finger bone discovered in a cave near Denisova in the Altai Mountains of Siberia. The DNA was well preserved because of the cold climate and a nearly-complete genomic sequence was obtained. Amazingly, Pääbo and his group found that the genome was that of an archaic hominin that diverged from the Neanderthals approximately 300,000 years ago, i.e., an eastern cousin of the Neanderthals. Even more amazingly, they found that between 4% and 6% of the genomes of present-day Melanesians is shared with the Denisovans. The most probable explanation is that the Denisovans ranged widely in Eastern Asia at the same time that the Neanderthals occupied the Western Eurasia and the Middle East, and that modern humans, on their way to Melanesia, interbred with them.

The hominid species shown in Table 6.1 show an overall progression in various characteristics: Their body size and brain size grew. They began to mature more slowly and to live longer. Their tools and weapons increased in sophistication. Meanwhile their teeth became smaller, and their skeletons more gracile — less heavy in proportion to their size. What were the evolutionary forces which produced these changes? How were they rewarded by a better chance of survival?

Our ancestors moved from a forest habitat to the savannas of Africa.

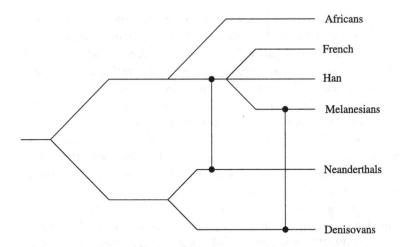

Fig. 6.1   This figure shows schematically the results of DNA sequencing by Svante Pääbo and his group at the Max Planck Institute for Evolutionary Anthropology. Their results indicate that modern humans outside Africa interbred with Neanderthals, and that Melanesians and the original people of Australia also interbred with the Denisovans, who were eastern cousins of the Neanderthals. Interbreeding is indicated by dots connected by vertical lines. (After Nature).

They changed from a vegetarian diet to an omnivorous one, becoming hunter-gatherers. The primate hand, evolved for grasping branches in a forest environment, found new uses. Branches and stones became weapons and tools — essential to hunters whose bodies lacked powerful claws and teeth. With a premium on skill in making tools, brain size increased. The beginnings of language helped to make hunts successful, and also helped in transmitting cultural skills, such as toolmaking and weaponmaking, from one generation to the next.

From the time scale shown in Tables 6.1 and 6.2, we can see that the coevolution of language, culture and intelligence took place over a period of several million years. As the cultures of the hominids became more complex, efficient transmission of skills and knowledge between generations required an increasingly complex language. This in turn required increased brain size and slow maturation, features which are built into the genomes of modern humans. A stable family structure and tribal social structure were also needed to protect the helpless offspring of our species as they slowly matured.

A modern human baby is almost entirely helpless. Compared with offspring of grazing animals, which are able to stand up and follow the herd immediately after birth, a human baby's development is almost ludicrously slow. However, there is nothing slow about the rate at which a young member of our species learns languages. Between the ages of one and four, young humans develop astonishing linguistic skills, far surpassing those of any other animal on earth. In the learning of languages by human children there is an interplay between genes and culture: The language learned is culturally determined, but the predisposition to learn some form of speech seems to be an inherited characteristic. For example, human babies of all nationalities have a tendency to "babble" — to produce random sounds. The sounds which they make are the same in all parts of the world, and they may include many sounds which are not used in the languages which the babies ultimately learn.

In his book, *Descent of Man* (John Murray, London, 1871) Charles Darwin wrote: "Man has an instinctive tendency to speak, as we see in the babble of young children, while no child has an instinctive tendency to bake, brew or write". Thus Darwin was aware of the genetic component of learning of speech by babies[2]. When our ancestors began to evolve a complex language and culture, it marked the start of an entirely new phase in the evolution of life on earth.

---

[2] Interestingly, a gene which seems to be closely associated with human speech has recently been located and mapped by C.S.L. Lai *et al*, who reported their results in Nature, **413**, 2001. These authors studied three generations of the "KE" family, 15 members of which are afflicted with a severe speech disorder. In all of the afflicted family members, a gene called FOXP2 on chromosome 7 is defective. In another unrelated individual, "CS", with a strikingly similar speech defect, the abnormality was produced by chromosomal translocation, the breakpoint coinciding exactly with the location of the FOXP2 gene. A still more recent study of the FOXP2 gene was published online in Nature AOP on August 14, 2002. The authors (Wolfgang Enard, Molly Przeworski, Cecilia S.L. Lai, Victor Wiebe, Takashi Kitano, Anthony P. Monaco, and Svante Pääbo) sequenced the FOXP2 gene and protein in the chimpanzee, gorilla, orangutan, rhesus macaque and mouse, comparing the results with sequences of human FOXP2. They found that in the line from the common ancestor of mouse and man to the point where the human genome branches away from that of the chimp, there are many nucleotide substitutions, but all are silent, i.e., they have no effect at all on the FOXP2 protein. The even more numerous non-silent DNA mutations which must have taken place during this period seem to have been rejected by natural selection because of the importance of conserving the form of the protein. However, in the human line after the human-chimp fork, something dramatic happens: There are only two base changes, but both of them affect the protein! This circumstance suggests to Enard et al that the two alterations in the human FOXP2 protein conferred a strong evolutionary advantage, and they speculate that this advantage may have been an improved capacity for language.

Y-chromosomal DNA and mitochondrial DNA

Recent DNA studies have cast much light on human prehistory, and especially on the story of how a small group of anatomically and behaviorally modern humans left Africa and populated the remainder of the world. Two types of DNA have been especially useful — Y-chromosomal DNA and mitochondrial DNA.

When we reproduce, the man's sperm carries either an X chromosome or a Y chromosome. It is almost equally probable which of the two it carries. The waiting egg of the mother has an X chromosome with complete certainty. When the sperm and egg unite to form a fertilized egg and later an embryo, the YX combinations become boys while the XX combinations become girls. Thus every male human carries a Y chromosome inherited from his father, and in fact this chromosome exists in every cell of a male's body.

Humans have a total of 23 chromosomes, and most of these participate in what might be called the "genetic lottery" — part of the remaining 22 chromosomes come from the father, and part from the mother, and it is a matter of chance which parent contributes which chromosome. Because of this genetic lottery, no two humans are genetically the same, except in the case of identical twins. This diversity is a great advantage, not only because it provides natural selection variation on which to act, but also it because prevents parasites from mimicking our cell-surface antigens and thus outwitting our immune systems. In fact the two advantages of diversity just mentioned are so great that sexual reproduction is almost universal among higher animals and plants.

Because of its special role in determining the sex of offspring, the Y chromosome is exempted from participation in the genetic lottery. This makes it an especially interesting object of study because the only changes that occur in Y chromosomes as they are handed down between generations are mutations. These mutations are not only infrequent but they also happen at a calculable rate. Thus by studying Y-chromosomal lineages, researchers have been able not only to build up prehistoric family trees but also to assign dates to events associated with the lineages.

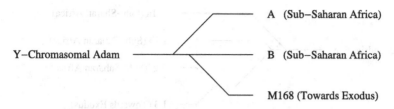

Fig. 6.2   The mutation M168 seems to have occurred just before the ancestral population of anatomically and behaviorally modern humans left Africa, roughly 70,000 years ago. All of the men who left Africa at that time carried this mutation. The descendents of this small group, probably a single tribe, were destined to populate the entire world outside Africa.

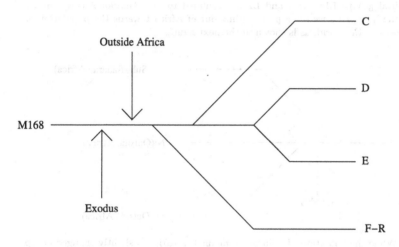

Fig. 6.3   After M168, further mutations occurred, giving rise to the Y-chromosomal groups C, D, E and F-R. Men carrying Y chromosomes of type C migrated to Central Asia, East Asia and Australia/New Guinea. The D group settled in Central Asia, while men carrying Y chromosomes of type E can be found today in East Asia, Sub-Saharan Africa, the Middle East, West Eurasia, and Central Asia. Populations carrying Y chromosomes of types F-R migrated to all parts of the world outside Africa. Those members of population P who found their way to the Americas carried the mutation M242. Only indigenous men of the Americas have Y chromosomes with M242.

Mitochondrial DNA is also exempted from participation in the genetic lottery, but for a different reason. In Chapter 3, we mentioned that mitochondria were once free-living eubacteria of a type called alphaproteobacteria. These free-living bacteria were able perform oxidative phosphorylation, i.e., they could couple the combustion of glucose to the formation of the high-energy phosphate bond in ATP. When photosynthesis

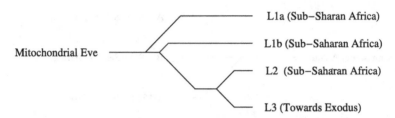

Fig. 6.4   Mitochondrial DNA is present in the bodies of both men and women, but is handed on only from mother to daughter. The human family tree constructed from mutations in mitochondrial DNA is closely parallel to the tree constructed by studying Y chromosomes. In both trees we see that only a single small group left Africa, and that the descendents of this small group populated the remainder of the world. The mitochondrial groups L1a, L1b, and L2 are confined to Sub-Saharan Africa, but by following the lineage L3 we see a path leading out of Africa towards the population of the remainder of the world, as is shown in the next figure.

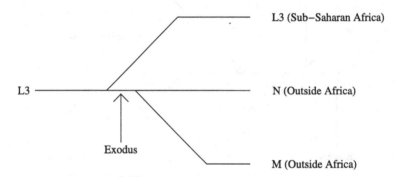

Fig. 6.5   While the unmutated L3 lineage remained in Africa, a slightly changed group of people found their way out. It seems to have been a surprisingly small group, perhaps only a single tribe. Their descendents populated the remainder of the the world. The branching between the N and M lineages occurred after their exodus from Africa. All women in Western Eurasia are daughters of the N line, while in Eastern Eurasia women are descended from both the N and M lineages. Daughters of both N and M reached the Americas.

evolved, the earth's atmosphere became rich in oxygen, which was a deadly poison to most of the organisms alive at the time. Two billion years ago, when atmospheric oxygen began to increase in earnest, many organisms retreated into anaerobic ecological niches, while others became extinct; but some survived the oxygen crisis by incorporating alpha-proteobacteria into their cells and living with them symbiotically. Today, mitochondria living as endosymbionts in all animal cells, use oxygen constructively to couple the burning of food with the synthesis of ATP. As a relic of the time when they were free-living bacteria, mitochondria have their own DNA, which is contained within them rather than within the cell nuclei.

When a sperm and an egg combine, the sperm's mitochondria are lost; and therefore all of the mitochondria in the body of a human child come from his or her mother. Just as Y-chromosomal DNA is passed essentially unchanged between generations in the male lines of a family tree, mitochondrial DNA is passed on almost without change in the female lines. The only changes in both cases are small and infrequent mutations. By estimating the frequency of these mutations, researchers can assign approximate dates to events in human prehistory.

Mitochondrial Eve and Y-Chromosomal Adam

On the female side of the human family tree, all lines lead back to a single woman, whom we might call "Mitochondrial Eve". Similarly, all the lines of the male family tree lead back to a single man, to whom we can give the name "Y-Chromosomal Adam". ("Eve" and "Adam" were not married, however; they were not even contemporaries!)

But why do the female and male and family trees both lead back to single individuals? This has to do with a phenomenon called "genetic drift". Sometimes a man will have no sons, and in that case, his male line will end, thus reducing the total number of Y-chromosomes in the population. Finally, after many generations, all Y-chromosomes will have dropped away through the ending of male lines except those that can be traced back to a single individual. Similar considerations hold for female lines.

When did Y-Chromosomal Adam walk the earth? Peter Underhill and his colleagues at Stanford University calculate on the basis of DNA evidence, that Adam lived between 40,000 and 140,000 years before the present (BP). However, on the basis of other evidence (for example the dating of archaeological sites in Australia) 40,000 years BP can be ruled out as being much too recent. Similar calculations on the date of Mitochondrial Eve find that she lived very approximately 150,000 years BP, but again there is a wide error range.

Exodus: Out of Africa

A model for the events leading up to the exodus of fully modern humans from Africa has been proposed by Sir Paul Mellars of Cambridge University, and it is shown in Table 6.3. In the article on which this table is

Table 6.3: Events leading up to the dispersal of fully modern humans from Africa (a model proposed by Sir Paul Mellars).

Years before present	Event
150,000–200,000 BP	Initial emergence of anatomically modern populations in Africa
110,000–90,000 BP	Temporary dispersal of anatomically modern populations (with Middle Paleolithic technology) from Africa to southwest Asia, associated with clear symbolic expression
80,000–70,000 BP	Rapid climatic and environmental changes in Africa
80,000–70,000 BP	Major technological, economic and social changes in south and east Africa
70,000–60,000 BP	Major population expansion in Africa from small source area
ca. 60,000 BP	Dispersal of modern populations from Africa to Eurasia

based, Mellars calls our attention to archaeological remains of anatomically modern humans at the sites of Skhul and Qafzeh in what is now northern Israel. The burials have been dated as having taken place 110,000–90,000 BP, and they show signs of cultural development, including ceremonial arrangement with arms folded, and sacrificial objects such as pierced shell ornaments. This early exodus was short-lived, however, probably because of competition with the long-established Neanderthal populations in the region.

In Mellars' model, rapid climatic and environmental changes took place in Africa during the period 80,000–70,000 BP. According to the Toba Catastrophe Theory[3] the climatic changes in Mellers' model were due to the eruption of a supervolcano at the site of what is now Lake Toba in Indonesia. This eruption, one of the largest known to us, took place ca. 73,000 BP, and plunged the earth into a decade of extreme cold, during which the population of our direct ancestors seem to have been reduced to a small number, perhaps as few as 10,000 individuals[4].

The survivors of the Toba Catastrophe may have been selected for improved linguistic ability, which gave them a more advanced culture than their contemporaries. Mellers points to archaeological and genetic evidence that a major population expansion of the L2 and L3 mitochondrial lineages took place in Africa 70,000–60,000 BP, starting from a small source region in East Africa, and spreading west and south. The expanding L2 and L3 populations were characterized by advanced cultural features such as upper paleolithic technology, painting and body ornaments.

All researchers agree that it was a small group of the L3 mitochondrial lineage that made the exodus from Africa ca. 70,000–60,000 BP, but there is some disagreement about the date of this event. These differences reflect the intrinsic inaccuracy of the genetic dating methods, but all experts in the field agree that the group passing out of Africa was remarkably small, especially when we reflect that the entire population of the remainder of the world is descended from them.

The men in this tiny but brave group of explorers carried with them the Y-chromosomal mutation M168, while the women were of the mitochondrial lineage L3. Shortly after they passed the Red Sea a mutation occurred and two new mitochondrial lineages were established, M and N. All women today in Western Eurasia are daughters of the N lineage[5], while the M lineage spread to the entire world outside Africa. The mitochondrial lineages M and N had further branches, and daughters of the A, B, C, D and X lineages passed over a land bridge which linked Siberia to Alaska during the period 22,000–7,000 BP, thus reaching the Americas.

In his excellent and fascinating book *Before the Dawn*, the science journalist Nicholas Wade discusses linguistic studies that support the early

---

[3]The Toba Catastrophe Theory is supported by such authors as Ann Gibons, Michael R. Rampino and Steven Self.

[4]Additional support to the Toba Catastrophe Theory comes from DNA studies of mammals, such as chimpanzees, orangutans, macques, cheetahs, tigers and gorillas. These mammals also seem, on the basis of DNA studies, to have been reduced to very small populations at the time of the Toba eruption.

[5]Of course, this broad statement does not take into account the movements of peoples that have taken place during historic times.

human migration scenarios that can be deduced from DNA research. The work of the unconventional but visionary linguist Joseph Greenberg of Stanford University is particularly interesting. While other linguists were content to demonstrate relationships between a few languages, such as those in the Indo-European family, Greenberg attempted to arrange all known languages into an enormous family tree. He published this work in the 1950's, long before the DNA studies that we have just been discussing, but because of what other linguists regarded as lack of rigor in his methods, Greenberg's prophetic voice was largely ignored by his peers. The linguist Paul Newman recalls visiting the London School of Oriental and African Studies ca. 1970. He was told that he could use the Common Room as long has he promised never to mention the name of Joseph Greenberg. Finally, after Joseph Greenberg's death, his visionary studies were vindicated by DNA-based human migration scenarios, which agreed in surprising detail with the great but neglected scholar's linguistically-based story of how early humans left their ancestral homeland in Africa and populated the entire earth.

Acceleration of human cultural evolution

In the caves of Spain and southern France are the remains of vigorous hunting cultures which flourished between 30,000 and 10,000 years ago. The people of these upper paleolithic cultures lived on the abundant cold-weather game which roamed the southern edge of the ice sheets during the Wurm glacial period: huge herds of reindeer, horses and wild cattle, as well as mammoths and wooly rhinos. The paintings found in the Dordogne region of France, for example, combine decorative and representational elements in a manner which contemporary artists might envy. Here and there among the paintings one can find stylized symbols which can be thought of as the first steps towards writing.

In this period, not only painting, but also tool-making and weapon-making were highly developed arts. For example, the Solutrian culture, which flourished in Spain and southern France about 20,000 years ago, produced beautifully worked stone lance points in the shape of laurel leaves and willow leaves. The appeal of these exquisitely pressure-flaked blades must have been aesthetic as well as functional. The people of the Solutrian culture had fine bone needles with eyes, bone and ivory pendants, beads and bracelets, and long bone pins with notches for arranging the hair. They also had red, yellow and black pigments for painting their bodies. The

Solutrian culture lasted for 4,000 years. It ended in about 17,000 B.C. when it was succeeded by the Magdalenian culture. Whether the Solutrian people were conquered by another migrating group of hunters, or whether they themselves developed the Magdalenian culture we do not know.

Beginning about 10,000 B.C., the way of life of the hunters was swept aside by a great cultural revolution: the invention of agriculture. The earth had entered a period of unusual climatic stability, and this may have helped to make agriculture possible. The first agricultural villages date from this time, as well as the earliest examples of pottery. Dogs and reindeer were domesticated, and later, sheep and goats. Radio-carbon dating shows that by 8,500 B.C., people living in the caves of Shanidar in the foothills of the Zagros mountains in Iran had domesticated sheep. By 7,000 B.C., the village farming community at Jarmo in Iraq had domesticated goats, together with barley and two different kinds of wheat.

Starting about 8000 B.C., rice came under cultivation in East Asia. This may represent an independent invention of agriculture, and agriculture may also have been invented independently in the western hemisphere, made possible by the earth's unusually stable climate during this period. At Jericho, in the Dead Sea valley, excavations have revealed a prepottery neolithic settlement surrounded by an impressive stone wall, six feet wide and twelve feet high. Radiocarbon dating shows that the defenses of the town were built about 7,000 B.C. Probably they represent the attempts of a settled agricultural people to defend themselves from the plundering raids of less advanced nomadic tribes.

Starting in western Asia, the neolithic agricultural revolution swept westward into Europe, and eastward into the regions that are now Iran and India. By 4,300 B.C., the agricultural revolution had spread southwest to the Nile valley, where excavations along the shore of Lake Fayum have revealed the remains of grain bins and silos. The Nile carried farming and stock-breeding techniques slowly southward, and wherever they arrived, they swept away the hunting and food-gathering cultures. By 3,200 B.C. the agricultural revolution had reached the Hyrax Hill site in Kenya. At this point the southward movement of agriculture was stopped by the swamps at the headwaters of the Nile. Meanwhile, the Mediterranean Sea and the Danube carried the revolution westward into Europe. Between 4,500 and 2,000 B.C. it spread across Europe as far as the British Isles and Scandinavia.

Early forms of writing

In Mesopotamia (which in Greek means "between the rivers"), the settled agricultural people of the Tigris and Euphrates valleys evolved a form of writing. The practical Mesopotamians seem to have invented writing as a means of keeping accounts.

Small clay and pebble counting tokens symbolizing items of trade began to be used in the Middle East about 9000 B.C., and they were widely used in the region until 1500 B.C. These tokens had various shapes, depending on the ware which they symbolized, and when made of clay they were often marked with parallel lines or crosses, which made their meaning more precise. In all, about 500 types of tokens have been found at various sites. Their use extended as far to the west as Khartoum in present-day Sudan, and as far to the east as the region which is now Pakistan. Often the tokens were kept in clay containers which were marked to indicate their contents. The markings on the containers, and the tokens themselves, evolved into true writing.

Among the earliest Mesopotamian writings are a set of clay tablets found at Tepe Yahya in southern Iran, the site of an ancient Elamite trading community halfway between Mesopotamia and India. The Elamite trade supplied the Sumerian civilization of Mesopotamia with silver, copper, tin, lead, precious gems, horses, timber, obsidian, alabaster and soapstone. The tablets found at Tepe Yahya are inscribed in proto-Elamite, and radiocarbon dating of organic remains associated with the tablets shows them to be from about 3,600 B.C. The inscriptions on these tablets were made by pressing the blunt and sharp ends of a stylus into soft clay. Similar tablets have been found at the Sumerian city of Susa at the head of the Tigris River.

In about 3,100 B.C. the cuneiform script was developed, and later Mesopotamian tablets are written in cuneiform, a phonetic script in which the symbols stand for syllables.

The Egyptian hieroglyphic (priest writing) system began its development in about 4,000 B.C. At that time, it was pictorial rather than phonetic. However, the Egyptians were in contact with the Sumerian civilization of Mesopotamia, and when the Sumerians developed a phonetic system of writing in about 3,100 B.C., the Egyptians were quick to adopt the idea. In the cuneiform writing of the Sumerians, a character stood for a syllable. In the Egyptian adaptation of this idea, some of the symbols stood for syllables or numbers, or were determinative symbols, which helped to make the meaning of a word more precise. However, some of the hieroglyphs were purely alphabetic, i.e., they stood for sounds which we would now represent by a single letter. This was important from the standpoint of cultural history, since it suggested to the Phoenicians the idea of an alphabet of the modern type.

In Sumer, the pictorial quality of the symbols was lost at a very early stage, so that in the cuneiform script the symbols are completely abstract. By contrast, the Egyptian system of writing was designed to decorate monuments and to be impressive even to an illiterate viewer; and this purpose was best served by retaining the elaborate pictographic form of the symbols. Besides the impressive and beautiful hieroglyphic writing which decorated their monuments, the Egyptians used a more rapidly-written script called Hieratic, which also existed in a shorthand version called Demotic (the "people's script"). During the Ptolemaic period[6] the Coptic language was spoken by part of the population of Egypt. This language was closely related to Egyptian, but was written using the Greek alphabet, an alphabet which developed from that of the Phoenicians. Knowledge of Coptic was one of the keys which helped Egyptologists to decipher the Rosetta stone, and hence the hieroglyphs[7].

Starting with the neolithic agricultural revolution and the invention of writing, human culture began to develop with explosive speed. Agriculture led to a settled way of life, with leisure for manufacturing complex artifacts, and for invention and experimentation. Writing allowed the cultural achievements of individuals or small groups to become widespread, and to be passed efficiently from one generation to the next.

Compared with the rate of ordinary genetic evolution, the speed with which the information-driven cultural evolution of Homo sapiens sapiens began to develop is truly astonishing. Twelve thousand years before the present, our ancestors were decorating the walls of their caves with drawings of mammoths. Only 10,000 years later, they were speculating about the existence of atoms! New methods for the conservation and utilization of information were the driving forces behind the explosively accelerating

---

[6] The first ruler in the Ptolomeic dynasty was one of the generals of Alexander the Great. The last in the line was the famous queen, Cleopatra, a contemporary (and lover) of Julius Caesar.

[7] It was the English physician, physicist and egyptologist Thomas Young who first argued that ancient Egyptian might be similar to Coptic. He was the first to decipher the Demotic script on the Rosetta stone, and he later compiled the first dictionary of Demotic Egyptian. He thus laid the foundation for Champollion's decoding of hieroglyphics in 1823.

evolution of human culture.

This remarkably rapid growth of human culture was not accompanied by very great genetic changes in our species. It took place instead because of a revolutionary leap in the efficiency with which information could be conserved and transmitted between generations, not in the code of DNA, but in the codes of Mesopotamian cuneiform, Egyptian hieroglyphics, Chinese ideograms, Mayan glyphs, and the Phoenecian and Greek alphabets.

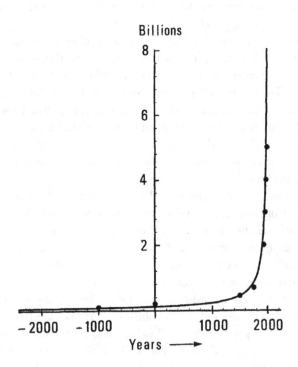

Fig. 6.6   Starting with the neolithic agricultural revolution and the invention of writing, human culture began to develop with explosive speed. This figure shows the estimated human population as a function of time during the last 4,000 years.   The dots are population estimates in billions, while the solid curve is the hyperbola $p = c/(2020 - y)$, where p is the global human population y is the year, and c = 234000. The curve reflects an explosively accelerating accumulation of information. Culturally transmitted techniques of agriculture allowed a much greater density of population than was possible for hunter-gatherers. The growth of population was further accelerated by the invention of printing and by the industrial and scientific developments which followed from this invention.

The invention of paper, ink, and printing

The ancient Egyptians were the first to make books. As early as 4,000 B.C., they began to make books in the form of scrolls by cutting papyrus reeds into thin strips and pasting them into sheets of double thickness. The sheets were glued together end to end, so that they formed a long roll. The rolls were sometimes very long indeed. For example, one roll, which is now in the British Museum, is 37 centimeters wide and 41 meters long.

The world's first great public library was established in Alexandria, Egypt, at the start of the Hellenistic Era (323 B.C. – 146 B.C.). Ptolemy I, who ruled Egypt after the dissolution of the empire of Alexander of Macedon, built a great library for the preservation of important manuscripts. The library at Alexandria was open to the general public, and at its height it was said to contain 750,000 volumes. Besides preserving important manuscripts, the library became a center for copying and distributing books.[8]

The material which the Alexandrian scribes used for making books was papyrus, which was relatively inexpensive. The Ptolemys were anxious that Egypt should keep its near-monopoly on book production, and they refused to permit the export of papyrus. Pergamum, a rival Hellenistic city in Asia Minor, also boasted a library second in size only to the great library at Alexandria. The scribes at Pergamum, unable to obtain papyrus from Egypt, tried to improve the preparation of the skins traditionally used for writing in Asia. The resulting material was called *Membranum pergamentum*, and in English, this name has become "parchment".

Paper of the type which we use today was not invented until 105 A.D. According to tradition, this enormously important invention was made by a Chinese eunuch named Tsai Lun. The kind of paper invented by Tsai Lun could be made from many things — for example, bark, hemp, rags, etc. The starting material was made into a pulp, mixed together with water and binder, spread out on a cloth to partially dry, and finally heated and pressed into thin sheets.

The Chinese later made another invention of immense importance to the cultural evolution of mankind. This was the invention of printing. Together with writing, printing is one of the key inventions which form the basis of human cultural evolution. The exact date of the invention of

---

[8] Unfortunately this great library was destroyed. Much damage was done in 145 B.C. by riots and civil war. According to some accounts, the destruction was completed in 47 B.C. when Julius Caesar's fleet set fire to the Egyptian fleet, and the fire spread to the city of Alexandria.

woodblock printing is uncertain, but indirect evidence makes it seem likely that the technique was first used in the Sui dynasty (581 A.D. – 618 A.D.). Woodblock printing became popular during the T'ang period (618 A.D. – 906 A.D.), and it was much used by Buddhist monks who were interested in producing many copies of the sacred texts which they had translated from Sanskrit. The act of reproducing prayers was also considered to be meritorious by the Buddhists.

Chinese administrators had for a long time followed the custom of brushing engraved official seals with ink and using them to stamp documents. The type of ink which they used was made from lampblack, water and binder. In fact, it was what we now call "India ink". However, in spite of its name, India ink is a Chinese invention, which later spread to India and from there to Europe.

We mentioned that paper of the type which we now use was invented in China in the first century A.D. Thus, the Buddhist monks of China had all the elements which they needed to make printing practical: They had good ink, cheap, smooth paper, and the tradition of stamping documents with ink-covered engraved seals. The first surviving block prints which they produced date from the 8th century A.D. They were made by carving a block of wood the size of a printed page so that raised characters remained, brushing ink onto the block, and pressing this onto a sheet of paper.

The oldest known printed book, the "Diamond Sutra", is dated 868 A.D., and it consists of only six printed pages. It was discovered in 1907 by an English scholar who obtained permission from Buddhist monks in Chinese Turkestan to open some walled-up monastery rooms, which were said to have been sealed for 900 years. The rooms were found to contain a library of about 15,000 manuscripts, among which was the Diamond Sutra. Block printing spread quickly throughout China, and also reached Japan, where woodblock printing ultimately reached great heights in the work of such artists as Hiroshige and Hokusai.

The invention of block printing during the T'ang dynasty had an enormously stimulating effect on literature, and the T'ang period is regarded as the golden age of Chinese lyric poetry. A collection of T'ang poetry, compiled in the 18th century, contains 48,900 poems by more than 2,000 poets.

About 1041–1048 A.D., a Chinese alchemist named Pi Sheng invented movable type, made from a mixture of clay and glue, and hardened by baking. He assembled the type into a text on an iron tray covered with a mixture of resin, wax and paper ash. He then gently heated the tray, and allowed it to cool, so that the type became firmly fixed in place. After printing as many copies of the text as he desired, Pi Sheng reheated the iron tray and reused the characters.

In 1313, a Chinese magistrate named Wang Chen initiated a large-scale printing project using movable type. He is said to have ordered craftsmen to carve 60,000 characters on movable wooden blocks. These were used to print a book on the history of technology. However, in spite of the efforts of Pi Sheng and Wang Chen, movable type never became very popular in China, because the Chinese written language contains 10,000 characters. However, printing with movable type was highly successful in Korea as early as the 15th century A.D., perhaps because a phonetic writing system existed in Korea, with symbols for syllables.

The unsuitability of the Chinese written language for the use of movable type was one of the greatest tragedies of Chinese civilization. Writing had been developed at a very early stage in Chinese history, but the system remained a pictographic system, with a different character for each word. A phonetic system of writing was never developed.

The failure to develop a phonetic system of writing had its roots in the Chinese imperial system of government. The Chinese empire formed a vast area in which many different languages were spoken. It was necessary to have a universal language of some kind in order to govern such an empire. The Chinese written language solved this problem admirably.

Suppose that the emperor sent identical letters to two officials in different districts. Reading the letters aloud, the officials might use entirely different words, although the characters in the letters were the same. Thus the Chinese written language was a sort of "Esperanto" which allowed communication between various language groups, and its usefulness as such prevented its replacement by a phonetic system.

The disadvantages of the Chinese system of writing were twofold: First, it was difficult to learn to read and write, and therefore literacy was confined to a small social class whose members could afford a prolonged education. The system of civil-service examinations made participation in the government dependent on a high degree of literacy, and hence the old, established scholar-gentry families maintained a long-term monopoly on power, wealth and education. Social mobility was possible in theory, since the civil service examinations were open to all, but in practice it was nearly unattainable.

The second great disadvantage of the Chinese system of writing was that it was unsuitable for printing with movable type. An "information explosion" occurred in the West following the introduction of printing with movable type, but this never occurred in China. It is ironical that although both paper and printing were invented by the Chinese, the full effect of these immensely important inventions bypassed China and instead revolutionized the west.

The information explosion

Like the process of silk manufacture, the art of papermaking remained for a long time a Chinese secret, but paper made in China (like silk) was traded with the Arab world along caravan routes. Finally, in 751, Chinese prisoners taken at the battle of Talas, near Samarkand, revealed the secret of papermaking to the Arabs. Between the 8th century and the 13th century, paper was extensively manufactured and used throughout the Islamic world, which stretched from the Middle East through North Africa to Spain. It seems strange that Chinese techniques of printing were not also transmitted during this period to the highly advanced Islamic civilization. Some historians believe that methods of printing were known to the Arabs in the 8th–13th centuries, but for religious reasons not used, the Koran being considered too holy to be reproduced by mechanical means. A further factor may have been the fact that the highly decorative classical Arabic script was not very well adapted to printing with movable type. Even in modern, simplified Arabic, each letter has many forms, whose use depends on the position in the word and on the neighboring letters.

Much of the knowledge achieved by the ancient civilizations of western Asia and the Mediterranean regions had been lost with the destruction of the great library of Alexandria. However, a few of the books of the classical and Hellenistic authors had survived in the eastern part of the Roman Empire at Byzantium.

The Byzantine empire included many Syriac-speaking subjects; and in fact, beginning in the 3rd century A.D., Syriac replaced Greek as the major language of western Asia. In the 5th century, there was a split in the Christian church of Byzantium, and the Nestorian church separated from the official Byzantine church. The Nestorians were bitterly persecuted, and therefore they migrated, first to Mesopotamia, and later to southwest Persia.

During the early Middle Ages, the Nestorian capital of Gondasapur was a great center of intellectual activity. The works of Plato, Aristotle, Hippocrates, Euclid, Archimedes, Ptolemy, Hero and Galen were translated into Syriac by Nestorian scholars, who had brought these books with them from Byzantium.

Among the most distinguished of the Nestorian translators were members of a family called Bukht-Yishu (meaning "Jesus hath delivered"), which produced seven generations of outstanding scholars. Members of this family were fluent not only in Greek and Syriac, but also in Arabic and Persian . In the 7th century, the Islamic religion suddenly emerged as a conquering and proselytizing force. The Arabs and their converts quickly conquered western Asia, northern Africa and Spain. After a short initial period of fanaticism which was often hostile to learning, the attitude of the Islamic conquerers changed to an appreciation of ancient cultures; and during the middle ages, the Islamic world reached a very high level of civilization[9]. Thus, while the century from 750 to 850 was primarily a period of translation from Greek into Syriac, the century from 850 to 950 was a period of translation from Syriac to Arabic.

The skill of the physicians of the Bukht-Yishu family convinced the Caliphs of Baghdad of the value of Greek learning, and in this way the family played an important role in the preservation of the classical cultures. Soon Baghdad replaced Gondasapur as a center of learning and translation.

Islamic scholars not only preserved our heritage from the ancient classical cultures but also added much to it. Chemistry, medicine, physics, astronomy and mathematics all owe much to the highly cultured Islamic world of the Middle Ages. The magnitude of this contribution can be judged from the many modern scientific words which have an Arabic origin. For example, the English words for chemistry is derived from the Arabic word "al-chimia", meaning "the changing". The word "al-kali", which appears in the writings of the Persian chemist Rahzes (860–950), means "the calcined" in Arabic. It is the source of our word "alkali" as well as of the symbol K for potassium. In mathematics, one of the most outstanding Arabic writers was al-Khwarizmi (780–850). The title of his book, *Ilm al-jabr wa'd muqabalah*, is the source of the English word "algebra". In Arabic, "al-jabr" means "the equating". Al-Khwarizmi's name has also become an English word, "algorism", the old word for arithmetic.

Towards the end of the Middle Ages, Europe began to be influenced by the advanced Islamic civilization. European scholars were anxious to learn, but there was an "iron curtain" of religious intolerance which made travel in the Islamic countries difficult and dangerous for Christians. However, in the 12th century, parts of Spain, including the city of Toledo, were reconquered

---

[9] There were, however, oscillations between periods of liberal fostering of intellectual efforts, and periods of puritanical suppression.

by the Christians. Toledo had been an Islamic cultural center, and many Moslem and Jewish scholars, together with their manuscripts, remained in the city when it passed into Christian hands. Thus Toledo became a center for the exchange of ideas between east and west; and it was in this city that many of the books of the classical Greek and Hellenistic philosophers were translated from Arabic into Latin.[10]

Another bridge between east and west was established by the Crusades. Crusaders returning from the Middle East brought paper with them to Europe, and from 1275 onwards the manufacture of paper became common in Italy, contributing importantly to the Italian Renaissance. In the 14th century, the manufacture of paper spread to France and Germany. Woodblock printing came into use in Europe during the last quarter of the 14th century. When the use of paper became common, it was noticed that the smooth and absorbent surface of paper was much more suitable for receiving a printed impression than was the surface of parchment, besides being far less expensive. In the 15th century, European artists such as Albrecht Dürer began to produce woodblock prints of great beauty. At the same time, woodblock printing was used to produce small books with a few pages of script, for example religious works and Latin grammars. Some experiments with movable wooden type seem to have been made in Holland, but the results were disappointing because when the letters were made as small as was desirable, they were not sufficiently durable.

Starting in approximately 1430, European craftsmen from the medieval guilds, who had a knowledge of the use of metal dies, began to apply this technique to printing. In the first step of this process, a set of dies, one for each letter of the alphabet, was engraved in brass or bronze. The dies were then used to produce a mold, over which lead was poured. When the lead plate was removed from the mold, the letters stood out in raised form. This method of metallographic printing was used in Holland and in the Rhineland, and in the period 1434–1439, Johannes Gutenberg used it in what is now Strasbourg France.

Gutenberg is generally credited with the simultaneous development (in 1450) of movable metal type and the printing press. He was a silversmith whose knowledge of metallurgy was undoubtedly useful to him when he

---

[10] Very often, the train of translations was very indirect, e.g., from Greek to Arabic to Hebrew to Spanish to Latin. For this reason, some of the earliest classical Greek texts made available to the Christian world were very incomplete. Added to this was the fact that translators and scribes felt quite free to edit, amend, and add to the texts on which they were working.

designed the machinery and type for printing. His partner in the book-producing enterprise was a businessman named Johann Fust. In 1509, there was a lawsuit in which Fust's grandson, Johann Schoffer, claimed that Fust alone was the inventor of the new printing method. However, in 1505, Schoffer had already written in a preface to an edition of Livy, "...the admirable art of typography was invented by the ingenious Johan Gutenberg at Mainz in 1450". One is more inclined to believe Schoffer's statement of 1505 than his later testimony in the 1509 lawsuit, which seems to have been motivated by the hope of financial gain.

The printing press invented by Gutenberg was a modification of the press which already used in Europe for binding books. The bed was fixed, and the movable upper platen (or level surface) was driven by a bar attached to a worm screw. The type letters were cast by pouring a molten alloy of lead, tin and antimony into moulds produced from dies. The letters were arranged into lines of type on wooden composing sticks held in the hand of the typographer, and each line was then justified (i.e., all the lines were made to have equal length) by the insertion of small blank lead spacers. After the printing of a page, each line was taken to pieces by hand, and the letters were returned to their containers.

Paper alone had an enormously stimulating effect on European culture when its manufacture and use became common at the end of the 13th century; but when paper was combined with Gutenberg's improved printing techniques in the 15th century, the combination produced an explosive accumulation of information. The combination of paper and improved printing resulted in the scientific and industrial revolutions, and in short the modern world.

One must add that it was not only paper and printing which combined to produce the information explosion, but also fragments from the writings of the classical ancient civilizations which had been translated first into Syriac, then from Syriac into Arabic, and finally from Arabic into Latin, and which thus, by a roundabout route, drifted into the consciousness of the west.

The career of Leonardo da Vinci illustrates the first phase of the information explosion: Inexpensive paper was being manufactured in Europe, and it formed the medium for Leonardo's thousands of pages of notes. His notes and sketches would never have been possible if he had been forced to use expensive parchment as a medium. On the other hand, the full force of Leonardo's genius and diligence was never felt because his notes were not printed. (In fact, fearing persecution for his radical ideas, Leonardo kept his notebooks secret.) Copernicus, who was a younger contemporary of Leonardo, had a much greater effect on the history of ideas because his work was published. Thus while paper alone made a large contribution to the information explosion, it was printing combined with paper which had an absolutely decisive and revolutionary impact: The modern scientific era began with the introduction of printing.

The development of printing in Europe and the rapid spread of books and knowledge produced a brilliant chainlike series of scientific discoveries — the sun-centered system of Copernicus, Kepler's three laws of planetary motion, Descartes' invention of analytic geometry, Gilbert's studies of magnetism, Galileo's discoveries in experimental physics and astronomy, the microscopy of Hooke and Leeuwenhoek, Newton's universal laws of motion and gravitation, the differential and integral calculus of Newton and Leibniz, the medical discoveries of Harvey, Jenner, Pasteur, Koch, Semmelweis and Lister, and the chemical discoveries of Boyle, Dalton, Priestly, Lavoisier and Berzelius.

The rapid accumulation of scientific knowledge made possible by paper and printing was quickly converted into the practical inventions of the industrial revolution. In the space of a few centuries, the information explosion changed Europe from a backward region into a society of an entirely new type, driven by scientific and technological innovation and by the diffusion and accumulation of knowledge.

Information-driven human cultural evolution as part of biological evolution

In thinking about human cultural evolution, one has a tendency to put it into a compartment by itself, separated from the evolution of microorganisms, animals, and plants. We feel that culture is not a subject for biologists but rather the domain of humanists. There is indeed a sharp qualitative discontinuity which marks the change from information transfer through the medium of DNA, RNA and proteins, to information transfer and accumulation through the medium of the spoken, written and printed word. Nevertheless it is important to remember that our species is a part of the biosphere, and that all our activities are fundamentally biological phenomena.

In Chapter 1, we discussed the ideas of Condorcet, one of the pioneers of evolutionary thought. He regarded genetic evolution (the process by which humans evolved from lower animals) and cultural evolution (the process by which civilized humans evolved from primitive man) as being two parts of a larger phenomenon which he called "progress". Although cultural evolution seems to differ qualitatively from genetic evolution, Condorcet regarded the two as being aspects of the same overall process.

Sharp qualitative discontinuities have occurred several times before during the earth's 4-billion year evolutionary history: A dramatic change occurred when autocatalytic systems first became surrounded by a cell membrane. Another sharp transition occurred when photosynthesis evolved, and a third when the enormously more complex eukaryotic cells developed from the prokaryotes. The evolution of multicellular organisms also represents a sharp qualitative change. Undoubtedly the change from molecular information transfer to cultural information transfer is an even more dramatic shift to a higher mode of evolution than the four sudden evolutionary gear-shifts just mentioned. Human cultural evolution began only an instant ago on the time-scale of genetic evolution. Already it has completely changed the planet. We have no idea where it will lead.

Suggestions for further reading

(1) Krause, J., et al., *The complete mitochondrial DNA genome of an unknown hominin from southern Siberia*, Nature **10**, 1038, (2010).

(2) Reich, D., et al., *Genetic history of an archaic hominin group from Denisova Cave in Siberia*, Nature **468**, 1053-60, (2010).

(3) Reich, D., et al., *Denisova Admixture and the First Modern Human Dispersals into Southeast Asia and Oceania*, Am. J. Human Genetics, **89**, 516-528, (2011).

(4) Wade, Nicholas, *Before the Dawn*, Penguin Books, (2006).

(5) Mellars, Paul, *The Emergence of Modern Humans*, Cornell University Press, Ithaca, N.Y., (1990).

(6) Mellars, Paul, *The Neanderthal Legacy*, Princeton University Press, Princeton, N.J., (1996).

(7) Mellars, Paul, *Why did modern human populations disperse from Africa ca. 60,000 years ago?*. Proceedings of the National Academy of Sciences **103 (25)**, 9381-6, (2006).

(8) Mellars, Paul, *Archeology and the Dispersal of Modern Humans in Europe: Deconstructing the "Aurignacian"*. Evolutionary Anthropology Issues News and Reviews **15 (5)**, 167, (2006).

(9) Mellars, Paul and Andrews, Martha V., *Excavations on Oronsay: Prehistoric Human Ecology on a Small Island*, Edinburgh University

Press, (1987).

(10) Forster, Peter, *Ice Ages and the mitochondrial DNA chronology of human dispersals: a review*, Phil. Trans. R. Soc. Lond. **B 359**, 255-264, (2004).

(11) Greenberg, J. H., Turner, C. G. and Zegura, L. Z., *The settlement of the Americas: a comparison of the linguistic, dental and genetic evidence*, Curr. Anthropol. **27**, 477-497, (1986)

(12) D.R. Griffin, *Animal Mind — Human Mind*, Dahlem Conferenzen 1982, Springer, Berlin, (1982).

(13) S. Savage-Rumbaugh, R. Lewin, et al., Kanzi: *The Ape at the Brink of the Human Mind*, John Wiley and Sons, New York, (1996).

(14) R. Dunbar, *Grooming, Gossip, and the Evolution of Language*, Harvard University Press, (1998).

(15) J.H. Greenberg, *Research on language universals*, Annual Review of Anthropology, **4**, 75-94 (1975).

(16) M.E. Bitterman, *The evolution of intelligence*, Scientific American, January, (1965).

(17) R. Fox, *In the beginning: Aspects of hominid behavioral evolution*, Man, **NS 2**, 415-433 (1967).

(18) M.S. Gazzaniga, *The split brain in man*, Scientific American, **217**, 24-29 (1967).

(19) D. Kimura, *The asymmetry of the human brain*, Scientific American, **228**, 70-78 (1973).

(20) R.G. Klein, *Anatomy, behavior, and modern human origins*, Journal of World Prehistory, **9 (2)**, 167-198 (1995).

(21) N.G. Jablonski and L.C. Aiello, editors, *The Origin and Diversification of Language*, Wattis Symposium Series in Anthropology. Memoirs of the California Academy of Sciences, **No. 24**, The California Academy of Sciences, San Francisco, (1998).

(22) S. Pinker, *The Language Instinct: How the Mind Creates Language*, Harper-Collins Publishers, New York, (1995).

(23) J.H. Barkow, L. Cosmides and J. Tooby, editors, *The Adapted Mind: Evolutionary Psychology and the Generation of Culture*, Oxford University Press, (1995).

(24) D.R. Begun, C.V. Ward and M.D. Rose, *Function, Phylogeny and Fossils: Miocene Hominid Evolution and Adaptations*, Plenum Press, New York, (1997).

(25) R.W. Byrne and A.W. Whitten, *Machiavellian Intelligence: Social Expertise and the Evolution of Intellect in Monkeys, Apes and Hu-*

mans, Cambridge University Press, (1988),

(26) V.P. Clark, P.A. Escholz and A.F. Rosa, editors, *Language: Readings in Language and Culture*, St Martin's Press, New York, (1997).

(27) T.W. Deacon, *The Symbolic Species: The Co-evolution of Language and the Brain*, W.W. Norton and Company, New York, (1997).

(28) C. Gamble, *Timewalkers: The Prehistory of Global Colonization*, Harvard University Press, (1994).

(29) K.R. Gibson and T. Inglod, editors, *Tools, Language and Cognition in Human Evolution*, Cambridge University Press, (1993).

(30) P. Mellers, *The Emergence of Modern Humans: An Archaeological Perspective*, Edinburgh University Press, (1990).

(31) P. Mellers, *The Neanderthal Legacy: An Archaeological Perspective of Western Europe*, Princeton University Press, (1996).

(32) S. Mithen, *The Prehistory of the Mind*, Thames and Hudson, London, (1996).

(33) D. Haraway, *Signs of dominance: from a physiology to a cybernetics, of primate biology*, C.R. Carpenter, 1939-1970, Studies in History of Biology, **6**, 129-219 (1983).

(34) D. Johanson and M. Edey, Lucy: *The Beginnings of Humankind*, Simon and Schuster, New York, (1981).

(35) B. Kurten, *Our Earliest Ancestors*, Colombia University Press, New York, (1992).

(36) R. Lass, *Historical Linguistics and Language Change*, Cambridge University Press, (1997).

(37) R.E. Leakey and R. Lewin, *Origins Reconsidered*, Doubleday, New York, (1992).

(38) P. Lieberman, *The Biology and Evolution of Language*, Harvard University Press, (1984).

(39) C.S.L. Lai, S.E. Fisher, J.A, Hurst, F. Vargha-Khadems, and A.P. Monaco, *A forkhead-domain gene is mutated in a severe speech and language disorder*, Nature, **413**, 519-523, (2001).

(40) W. Enard, M. Przeworski, S.E. Fisher, C.S.L. Lai, V. Wiebe, T. Kitano, A.P. Monaco, and S. Paabo, *Molecular evolution of FOXP2, a gene involved in speech and language*, Nature AOP, published online 14 August 2002.

(41) M. Gopnik and M.B. Crago, *Familial aggregation of a developmental language disorder*, Cognition, **39**, 1-50 (1991).

(42) K.E. Watkins, N.F. Dronkers, and F. Vargha-Khadem, *Behavioural analysis of an inherited speech and language disorder. Comparison*

with acquired aphasia, Brain, **125**, 452-464 (2002).

(43) J.D. Wall and M. Przeworski, *When did the human population size start increasing?*, Genetics, **155**, 1865-1874 (2000).

(44) L. Aiello and C. Dean, An *Introduction to Human Evolutionary Anatomy*, Academic Press, London, (1990).

(45) F. Ikawa-Smith, ed., *Early Paleolithic in South and East Asia*, Mouton, The Hague, (1978).

(46) M. Aitken, *Science Based Dating in Archeology*, Longman, London, (1990).

(47) R.R. Baker, *Migration: Paths Through Space and Time*, Hodder and Stoughton, London, (1982).

(48) P. Bellwood, *Prehistory of the Indo-Malaysian Archipelago*, Academic Press, Sidney, (1985).

(49) P.J. Bowler, *Theories of Human Evolution: A Century of Debate*, 1884-1944, Basil Blackwell, Oxford, (1986).

(50) G. Isaac and M. McCown, eds., *Human Origins: Louis Leaky and the East African Evidence*, Benjamin, Menlo Park, (1976).

(51) F.J. Brown, R. Leaky, and A. Walker, *Early Homo erectus skeleton from west Lake Turkana, Kenya*, Nature, **316**, 788-92 (1985).

(52) K.W. Butzer, *Archeology as Human Ecology*, Cambridge University Press, (1982).

(53) A.T. Chamberlain and B.A. Wood, *Early hominid phylogeny*, Journal of Human Evolution, **16**, 119-33, (1987).

(54) P. Mellars and C. Stringer, eds., *The Human Revolution: Behavioural and Biological Perspectives in the Origins of Modern Humans*, Edinburgh University Press, (1989).

(55) G.C. Conroy, *Primate Evolution*, W.W. Norton, New York, (1990).

(56) R.I.M. Dunbar, *Primate Social Systems*, Croom Helm, London, (1988).

(57) B. Fagan, *The Great Journey: The Peopling of Ancient America*, Thames and Hudson, London, (1987).

(58) R.A. Foley, ed., *Hominid Evolution and Community Ecology*, Academic Press, New York, (1984).

(59) S.R. Binford and L.R. Binford, *Stone tools and human behavior*, Scientific American, **220**, 70-84, (1969).

(60) G. Klein, *The Human Career, Human Biological and Cultural Origins*, University of Chicago Press, (1989).

(61) B.F. Skinner and N. Chomsky, *Verbal behavior, Language*, **35**, 26-58 (1959).

(62) D. Bickerton, *The Roots of Language*, Karoma, Ann Arbor, Mich., (1981).

(63) E. Lenneberg in *The Structure of Language: Readings in the Philosophy of Language*, J.A. Fodor and J.A. Katz editors, Prentice-Hall, Englewood Cliffs N.J., (1964).

(64) S. Pinker, Talk of genetics and visa versa, Nature, **413**, 465-466, (2001).

(65) S. Pinker, *Words and rules in the human brain*, Nature, **387**, 547-548, (1997).

(66) M. Ruhelen, *The Origin of Language*, Wiley, New York, (1994).

(67) C.B. Stringer and R. McKie, *African Exodus: The Origins of Modern Humanity*, Johnathan Cape, London (1996).

(68) R. Lee and I. DeVore, editors, *Kalahari Hunter-Gatherers*, Harvard University Press, (1975).

(69) R.W. Sussman, *The Biological Basis of Human Behavior*, Prentice Hall, Englewood Cliffs, (1997).

(70) D. Schamand-Besserat, *Before Writing, Volume 1, From Counting to Cuneiform*, University of Texas Press, Austin, (1992).

(71) D. Schmandnt-Besserat, *How Writing Came About*, University of Texas Press, Austin, (1992).

(72) A. Robinson, *The Story of Writing*, Thames, London, (1995).

(73) A. Robinson, *Lost Languages: The Enegma of the World's Great Undeciphered Scripts*, McGraw-Hill, (2002).

(74) D. Jackson, *The Story of Writing*, Taplinger, New York, (1981).

(75) G. Jeans, *Writing: The Story of Alphabets and Scripts*, Abrams and Thames, (1992).

(76) W.M. Senner, editor, *The Origins of Writing*, University of Nebraska Press, Lincoln and London, (1989).

(77) F. Coulmas, *The Writing Systems of the World*, Blackwell, Oxford, (1989).

(78) F. Coulmas, *The Blackwell Encyclopedia of Writing Systems*, Blackwell, Oxford, (1996).

(79) P.T. Daniels and W. Bright, editors, *The World's Writing Systems*, Oxford University Press, (1996).

(80) H.J. Nissen, *The Early History of the Ancient Near East*, 9000-2000 B.C., University of Chicago Press, (1988).

(81) H.J. Nissen, *Archaic Bookkeeping: Early Writing and Techniques of Economic Administration in the Ancient Near East*, University of Chicago Press, (1993).

(82) J. Bottero, *Ancient Mesopotamia: Everyday Life in the First Civilization*, Edinburgh University Press, (2001).

(83) J. Bottero, *Mesopotamia: Writing, Reasoning and the Gods*, University of Chicago Press, (1992).

(84) J.T. Hooker, *Reading the Past: Ancient Writing, from Cuneiform to the Alphabet*, University of California Press, Berkeley and Los Angeles, (1990).

(85) W.A. Fairservis, Jr., *The Script of the Indus Valley*, Scientific American, March (1983), 41-49.

(86) C.H. Gordon, *Forgotten Scripts: Their Ongoing Discovery and Decipherment*, Dorset Press, New York, (1992).

(87) G. Ferraro, *Cultural Anthropology, 3rd Edition*, Wadsworth, Belmont CA, (1998).

(88) R. David, *Handbook to Life in Ancient Egypt*, Facts on File, New York, (1998).

(89) D. Sandison, *The Art of Egyptian Hieroglyphs*, Reed, London, (1997).

(90) K.T. Zauzich, *Hieroglyphs Without Mystery*, University of Texas Press, Austin, (1992).

(91) B. Watterson, *Introducing Egyptian Hieroglyphs*, Scottish Academic Press, Edinburgh, (1981).

(92) M. Pope, *The Story of Decipherment, from Egyptian Hieroglyphs to Maya Script*, Thames and Hudson, London, (1999).

(93) M.D. Coe, *Breaking the Maya Script*, Thames and Hudson, New York, (1992).

(94) M.D. Coe, *The Maya, 5th Edition*, Thames and Hudson, New York, (1993).

(95) M.D. Coe, *Mexico: From the Olmecs to the Aztecs, 4th Edition*, Thames and Hudson, New York, (1994).

(96) D. Preidel, L. Schele and J. Parker, *Maya Cosmos: Three Thousand Years on the Shaman's Path*, William Morrow, New York, (1993).

(97) W.G. Bolz, *The Origin and Early Development of the Chinese Writing System*, American Oriental Society, New Haven Conn., (1994).

(98) T.F. Carter, *The Invention of Printing in China and its Spread Westward*, Ronald Press, (1925).

(99) E. Eisenstein, *The Printing Revolution in Early Modern Europe*, Cambridge University Press, (1983).

(100) M. Olmert, *The Smithsonian Book of Books*, Wing Books, New York, (1992).

Chapter 7

INFORMATION TECHNOLOGY

The first computers

If civilization survives, historians in the distant future will undoubtedly regard the invention of computers as one of the most important steps in human cultural evolution — as important as the invention of writing or the invention of printing. The possibilities of artificial intelligence have barely begun to be explored, but already the impact of computers on society is enormous.

The first programmable universal computers were completed in the mid-1940's; but they had their roots in the much earlier ideas of Blaise Pascal (1623–1662), Gottfried Wilhelm Leibniz (1646–1716), Joseph Marie Jacquard (1752–1834) and Charles Babbage (1791–1871).

In 1642, the distinguished French mathematician and philosopher Blaise Pascal completed a working model of a machine for adding and subtracting. According to tradition, the idea for his "calculating box" came to Pascal when, as a young man of 17, he sat thinking of ways to help his father (who was a tax collector). In describing his machine, Pascal wrote: "I submit to the public a small machine of my own invention, by means of which you alone may, without any effort, perform all the operations of arithmetic, and may be relieved of the work which has often times fatigued your spirit when you have worked with the counters or with the pen".

Pascal's machine worked by means of toothed wheels. It was much improved by Leibniz, who constructed a mechanical calculator which, besides adding and subtracting, could also multiply and divide. His first machine was completed in 1671; and Leibniz' description of it, written in Latin, is preserved in the Royal Library at Hanover: "There are two parts of the machine, one designed for addition (and subtraction), and the other designed for multiplication (and division); and they should fit together. The adding (and subtracting) machine coincides completely with the calculating box of Pascal. Something, however, must be added for the sake of multiplication..."

"The wheels which represent the multiplicand are all of the same size, equal to that of the wheels of addition, and are also provided with ten teeth which, however, are movable so that at one time there should protrude 5, at another 6 teeth, etc., according to whether the multiplicand is to be represented five times or six times, etc."

"For example, the multiplicand 365 consists of three digits, 3, 6, and 5. Hence the same number of wheels is to be used. On these wheels, the multiplicand will be set if from the right wheel there protrude 5 teeth, from the middle wheel 6, and from the left wheel 3".

By 1810, calculating machines based on Leibniz' design were being manufactured commercially; and mechanical calculators of a similar (if much improved) design could be found in laboratories and offices until the 1960's. The idea of a programmable universal computer is due to the English mathematician, Charles Babbage, who was the Lucasian Professor of Mathematics at Cambridge University. (In the 17th century, Isaac Newton held this post, and in the 20th century, P.A.M. Dirac and Stephen Hawking also held it.)

In 1812, Babbage conceived the idea of constructing a machine which could automatically produce tables of functions, provided that the functions could be approximated by polynomials. He constructed a small machine, which was able to calculate tables of quadratic functions to eight decimal places, and in 1832 he demonstrated this machine to the Royal Society and to representatives of the British government.

The demonstration was so successful that Babbage secured financial support for the construction of a large machine which would tabulate sixth-order polynomials to twenty decimal places. The large machine was never completed, and twenty years later, after having spent seventeen thousand pounds on the project, the British government withdrew its support. The reason why Babbage's large machine was never finished can be understood from the following account by Lord Moulton of a visit to the mathematician's laboratory:

"One of the sad memories of my life is a visit to the celebrated mathematician and inventor, Mr. Babbage. He was far advanced in age, but his mind was still as vigorous as ever. He took me through his workrooms".

"In the first room I saw the parts of the original Calculating Machine, which had been shown in an incomplete state many years before, and had even been put to some use. I asked him about its present form. 'I have not finished it, because in working at it, I came on the idea of my Analytical Machine, which would do all that it was capable of doing, and much more. Indeed, the idea was so much simpler that it would have taken more work to complete the Calculating Machine than to design and construct the other in its entirety; so I turned my attention to the Analytical Machine'".

"After a few minutes talk, we went into the next workroom, where he showed me the working of the elements of the Analytical Machine. I asked if I could see it. 'I have never completed it,' he said, 'because I hit upon the idea of doing the same thing by a different and far more effective method, and this rendered it useless to proceed on the old lines.'"

"Then we went into a third room. There lay scattered bits of mechanism, but I saw no trace of any working machine. Very cautiously I approached the subject, and received the dreaded answer: 'It is not constructed yet, but I am working at it, and will take less time to construct it altogether than it would have taken to complete the Analytical Machine from the stage in which I left it.' I took leave of the old man with a heavy heart".

Babbage's first calculating machine was a special-purpose mechanical computer, designed to tabulate polynomial functions; and he abandoned this design because he had hit on the idea of a universal programmable computer. Several years earlier, the French inventor Joseph Marie Jacquard had constructed an automatic loom in which large wooden "punched cards" were used to control the warp threads. Inspired by Jacquard's invention, Babbage planned to use punched cards to program his universal computer. (Jacquard's looms could be programmed to weave extremely complex patterns: A portrait of the inventor, woven on one of his looms in Lyon, hung in Babbage's drawing room.)

One of Babbage's frequent visitors was Augusta Ada[1], Countess of Lovelace (1815–1852), the daughter of Lord and Lady Byron. She was a mathematician of considerable ability, and it is through her lucid descriptions that we know how Babbage's never-completed Analytical Machine was to have worked.

The next step towards modern computers was taken by Herman Hollerith, a statistician working for the United States Bureau of the Census. He invented electromechanical machines for reading and sorting data punched onto cards. Hollerith's machines were used to analyze the data from the 1890 United States Census. Because the Census Bureau was a very limited

---

[1] The programming language ADA is named after her.

market, Hollerith branched out and began to manufacture similar machines for use in business and administration. His company was later bought out by Thomas J. Watson, who changed its name to International Business Machines.

In 1937, Howard Aiken, of Harvard University, became interested in combining Babbage's ideas with some of the techniques which had developed from Hollerith's punched card machines. He approached the International Business Machine Corporation, the largest manufacturer of punched card equipment, with a proposal for the construction of a large, automatic, programmable calculating machine.

Aiken's machine, the Automatic Sequence Controlled Calculator (ASCC), was completed in 1944 and presented to Harvard University. Based on geared wheels, in the Pascal-Leibniz-Babbage tradition, ASCC had more than three quarters of a million parts and used 500 miles of wire. ASCC was unbelievably slow by modern standards — it took three-tenths of a second to perform an addition — but it was one of the first programmable general-purpose digital computers ever completed. It remained in continuous use, day and night, for fifteen years.

In the ASCC, binary numbers were represented by relays, which could be either on or off. The on position represented 1, while the off position represented 0, these being the only two digits required to represent numbers in the binary (base 2) system. Electromechanical calculators similar to ASCC were developed independently by Konrad Zuse in Germany and by George R. Stibitz at the Bell Telephone Laboratory.

Electronic digital computers

In 1937, the English mathematician A.M. Turing published an important article in the Proceedings of the London Mathematical Society in which envisioned a type of calculating machine consisting of a long row of cells (the "tape"), a reading and writing head, and a set of instructions specifying the way in which the head should move the tape and modify the state and "color" of the cells on the tape. According to a hypothesis which came to be known as the "Church-Turing hypothesis", the type of computer proposed by Turing was capable of performing every possible type of calculation. In other words, the Turing machine could function as a universal computer.

In 1943, a group of English engineers, inspired by the ideas of Alan Turing and those of the mathematician M.H.A. Newman, completed the electronic digital computer Colossus. Colossus was the first large-scale electronic computer. It was used to break the German Enigma code; and it thus affected the course of World War II.

In 1946, ENIAC (Electronic Numerical Integrator and Calculator) became operational. This general-purpose computer, designed by J.P. Eckert and J.W. Mauchley of the University of Pennsylvania, contained 18,000 vacuum tubes, one or another of which was often out of order. However, during the periods when all its vacuum tubes were working, an electronic computer like Colossus or ENIAC could shoot ahead of an electromechanical machine (such as ASCC) like a hare outdistancing a tortoise.

During the summer of 1946, a course on "The Theory and Techniques of Electronic Digital Computers" was given at the University of Pennsylvania. The ideas put forward in this course had been worked out by a group of mathematicians and engineers headed by J.P. Eckert, J.W. Mauchley and John von Neumann, and these ideas very much influenced all subsequent computer design.

Cybernetics

The word "Cybernetics", was coined by the American mathematician Norbert Wiener (1894–1964) and his colleagues, who defined it as "the entire field of control and communication theory, whether in the machine or in the animal". Wiener derived the word from the Greek term for "steersman".

Norbert Wiener began life as a child prodigy: He entered Tufts University at the age of 11 and received his Ph.D. from Harvard at 19. He later became a professor of mathematics at the Massachusetts Institute of Technology. In 1940, with war on the horizon, Wiener sent a memorandum to Vannevar Bush, another MIT professor who had done pioneering work with analogue computers, and had afterwards become the chairman of the U.S. National Defense Research Committee. Wiener's memorandum urged the American government to support the design and construction of electronic digital computers, which would make use of binary numbers, vacuum tubes, and rapid memories. In such machines, the memorandum emphasized, no human intervention should be required except when data was to be read into or out of the machine.

Like Leo Szilard, John von Neumann, Claude Shannon and Erwin Schrödinger, Norbert Wiener was aware of the relation between information and entropy. In his 1948 book Cybernetics he wrote: "...we had to develop a statistical theory of the amount of information, in which the unit amount of information was that transmitted by a single decision between equally probable alternatives. This idea occurred at about the same time to several writers, among them the statistician R.A. Fisher, Dr. Shannon of Bell Telephone Laboratories, and the author. Fisher's motive in studying this subject is to be found in classical statistical theory; that of Shannon in the problem of coding information; and that of the author in the problem of noise and message in electrical filters... The notion of the amount of information attaches itself very naturally to a classical notion in statistical mechanics: that of entropy. Just as the amount of information in a system is a measure of its degree of organization, so the entropy of a system is a measure of its degree of disorganization; and the one is simply the negative of the other".

During World War II, Norbert Wiener developed automatic systems for control of anti-aircraft guns. His systems made use of feedback loops closely analogous to those with which animals coordinate their movements. In the early 1940's, he was invited to attend a series of monthly dinner parties organized by Arturo Rosenblueth, a professor of physiology at Harvard University. The purpose of these dinners was to promote discussions and collaborations between scientists belonging to different disciplines. The discussions which took place at these dinners made both Wiener and Rosenblueth aware of the relatedness of a set of problems that included homeostasis and feedback in biology, communication and control mechanisms in neurophysiology, social communication among animals (or humans), and control and communication involving machines.

Wiener and Rosenblueth therefore tried to bring together workers in the relevant fields to try to develop common terminology and methods. Among the many people whom they contacted were the anthropologists Gregory Bateson and Margaret Mead, Howard Aiken (the designer of the Automatic Sequence Controlled Calculator), and the mathematician John von Neumann. The Josiah Macy Jr. Foundation sponsored a series of ten yearly meetings, which continued until 1949 and which established cybernetics as a new research discipline. It united areas of mathematics, engineering, biology, and sociology which had previously been considered unrelated. Among the most important participants (in addition to Wiener, Rosenblueth, Bateson, Mead, and von Neumann) were Heinz von Foerster, Kurt Lewin, Warren McCulloch and Walter Pitts. The Macy conferences were small and informal, with an emphasis on discussion as opposed to the presentation of formal papers. A stenographic record of the last five conferences has been published, edited by von Foerster. Transcripts of the discussions give a vivid picture of the enthusiastic and creative atmosphere of the meetings. The participants at the Macy Conferences perceived Cybernetics as a much-needed bridge between the natural sciences and the humanities. Hence their enthusiasm. Wiener's feedback loops and von Neumann's theory of games were used by anthropologists Mead and Bateson to explain many aspects of human behavior.

Microelectronics

The problem of unreliable vacuum tubes was solved in 1948 by John Bardeen, William Shockley and Walter Brattain of the Bell Telephone Laboratories. Application of quantum theory to solids had led to an understanding of the electronic properties of crystals. Like atoms, crystals were found to have allowed and forbidden energy levels.

The allowed energy levels for an electron in a crystal were known to form bands; i.e., some energy ranges with a quasi-continuum of allowed states (allowed bands), and other energy ranges with none (forbidden bands). The lowest allowed bands were occupied by electrons, while higher bands were empty. The highest filled band was called the valence band, and the lowest empty band was called the conduction band.

According to quantum theory, whenever the valence band of a crystal is only partly filled, the crystal is a conductor of electricity; but if the valence band is completely filled with electrons, the crystal is an electrical insulator. (A completely filled band is analogous to a room so packed with people that none of them can move.)

In addition to explaining conductors and insulators, quantum theory yielded an understanding of semiconductors — crystals where the valence band is completely filled with electrons, but where the energy gap between the conduction band and the valence band is relatively small. For example, crystals of the elements silicon and germanium are semiconductors. For such a crystal, thermal energy is sometimes enough to lift an electron from the valence band to the conduction band.

During the period from 1947 to 1951, John Bardeen and Walter Houser Brattain, Shockley co-invented the transistor, for which all three were awarded the 1956 Nobel Prize in physics. Working at the Bell Telephone Laboratories, they found ways to control the conductivity of germanium crystals by injecting electrons into the conduction band, or alternatively by removing electrons from the valence band. They could do this by forming junctions between crystals "doped" with appropriate impurities, and by injecting electrons with a special electrode. The semi-conducting crystals whose conductivity was controlled in this way could be used as electronic valves, in place of vacuum tubes.

By the 1960's, replacement of vacuum tubes by transistors in electronic computers had led not only to an enormous increase in reliability and a great reduction in cost, but also to an enormous increase in speed. It was found that the limiting factor in computer speed was the time needed for an electrical signal to propagate from one part of the central processing unit to another. Since electrical impulses propagate with the speed of light, this time is extremely small; but nevertheless, it is the limiting factor in the speed of electronic computers.

In order to reduce the propagation time, computer designers tried to make the central processing units very small; and the result was the development of integrated circuits and microelectronics. (Another motive for miniaturization of electronics came from the requirements of space exploration.)

Integrated circuits were developed[2], in which single circuit elements were not manufactured separately, but instead the whole circuit was made at one time. An integrated circuit is a multilayer sandwich-like structure, with conducting, resisting and insulating layers interspersed with layers of germanium or silicon, "doped" with appropriate impurities. At the start of the manufacturing process, an engineer makes a large drawing of each layer. For example, the drawing of a conducting layer would contain pathways which fill the role played by wires in a conventional circuit, while the remainder of the layer would consist of areas destined to be etched away by acid.

The next step is to reduce the size of the drawing and to multiply it photographically. The pattern of the layer is thus repeated many times, like the design on a piece of wallpaper. The multiplied and reduced drawing is then focused through a reversed microscope onto the surface to be etched.

---

[2]The story of the development of integrated circuits is a colorful one: In 1956, a branch of the Beckman Instruments was established at 391 San Antonio Road, Mountain View, California, under Shockley's direction. Its purpose was to develop transistors commercially. Unable to persuade his Bell colleagues to move to California, Shockley recruited an entirely new team. However, after experiencing Shockley's autocratic and abrasive management style for a year, the team resigned *en masse* in 1957 and formed a new company, Fairchild Semiconductor. The "traitorous eight" who resigned from Shockley's laboratory included Robert Noyce and Gordon Moore, who later also founded Intel, and Jean Hoerni, who introduced the planar process for making transistors. Together with Jack Kilby, Noyce is credited with the development of the first practical integrated circuits. Almost all of the companies that now occupy California's "Silicon Valley" owe their origin to Fairchild Semiconductor and the "traitorous eight".

Successive layers are built up by evaporating or depositing thin films of the appropriate substances onto the surface of a silicon or germanium wafer. If the layer being made is to be conducting, the surface might consist of an extremely thin layer of copper, covered with a photosensitive layer called a "photoresist". On those portions of the surface receiving light from the pattern, the photoresist becomes insoluble, while on those areas not receiving light, the photoresist can be washed away.

The surface is then etched with acid, which removes the copper from those areas not protected by photoresist. Each successive layer of a wafer is made in this way, and finally the wafer is cut into tiny "chips", each of which corresponds to one unit of the wallpaper-like pattern. Although the area of a chip may be much smaller than a square centimeter, the chip can contain an extremely complex circuit.

In 1965, only four years after the first integrated circuits had been produced, Dr. Gordon E. Moore, one of the founders of Intel, made a famous prediction which has come to be known as "Moore's Law". He predicted that the number of transistors per integrated circuit would double every two years, and that this trend would continue through 1975. In fact, the general trend predicted by Moore has continued for a much longer time. Although the number of transistors per unit area has not continued to double every two years, the logic density (bits per unit area) has done so, and a modified version of Moore's law continues to hold. How much longer the trend can continue remains to be seen. Physical limits to miniaturization of transistors of the present type will soon be reached; but there is hope that further miniaturization can be achieved through "quantum dot" technology, molecular switches, and autoassembly, as will be discussed in the next chapter.

In 1971, 2,300 transistors were being placed on a single integrated circuit chip. A typical programmable minicomputer or "microprocessor", manufactured in the mid-1970's, could have 30,000 circuit elements, all of which were contained on a single chip. By 1989, more than a million transistors were being placed on a chip; and by 2000, the number reached 42,000,000. By 2011, the number of transistors per chip had increased to a staggering 2,600,000,000.

As a result of miniaturization and parallelization, the speed of computers rose exponentially. In 1960, the fastest computers could perform a hundred thousand elementary operations in a second. By 1970, the fastest computers took less than a second to perform a million such operations. In 1987, a massively parallel computer, with 566 parallel processors, called

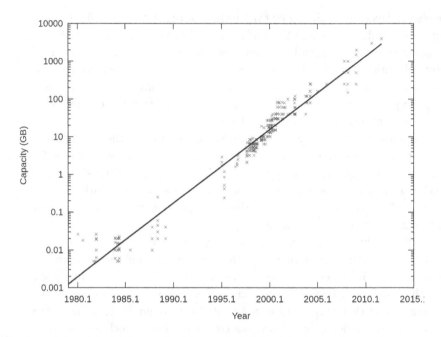

Fig. 7.1 A logarithmic plot of the increase in PC hard-drive capacity in gigabytes. An extrapolation of the rate of increase predicts that the capacity of an individual, commercially available PC will reach 10,000 gigabytes by 2015, i.e., 10,000,000,000,000 bytes. (After Hankwang and Rentar, Wikimedia Commons.)

GFll was designed to perform 11 billion floating-point operations per second (flops). By 2002, the fastest computer performed 40 at teraflops, making use of 5120 parallel CPU's.

Computer disk storage has also undergone a remarkable development. In 1987, the magnetic disk storage being produced could store 20 million bits of information per square inch; and even higher densities could be achieved by optical storage devices. Storage density has until followed a law similar to Moore's law, as is shown in Figure 7.1.

In the 1970's and 1980's, computer networks were set up linking machines in various parts of the world. It became possible (for example) for a scientist in Europe to perform a calculation interactively on a computer in the United States just as though the distant machine were in the same room; and two or more computers could be linked for performing large calculations. It also became possible to exchange programs, data, letters and manuscripts very rapidly through the computer networks.

The exchange of large quantities of information through computer networks was made easier by the introduction of fiber optics cables. By 1986, 250,000 miles of such cables had been installed in the United States. If a ray of light, propagating in a medium with a large refractive index, strikes the surface of the medium at a grazing angle, then the ray undergoes total internal reflection. This phenomenon is utilized in fiber optics: A light signal can propagate through a long, hairlike glass fiber, following the bends of the fiber without losing intensity because of total internal reflection. However, before fiber optics could be used for information transmission over long distances, a technological breakthrough in glass manufacture was needed, since the clearest glass available in 1940 was opaque in lengths more than 10 m. Through studies of the microscopic properties of glasses, the problem of absorption was overcome. By 1987, devices were being manufactured commercially that were capable of transmitting information through fiber-optic cables at the rate of 1.7 billion bits per second.

The history of the Internet and World Wide Web

The history of the Internet began in 1961, when Leonard Kleinrock, a student at MIT, submitted a proposal for Ph.D. thesis entitled "Information Flow in Large Communication Nets". In his statement of the problem, Kleinrock wrote: "The nets under consideration consist of nodes, connected to each other by links. The nodes receive, sort, store, and transmit messages that enter and leave via the links. The links consist of one-way channels, with fixed capacities. Among the typical systems which fit this description are the Post Office System, telegraph systems, and satellite communication systems." Kleinrock's theoretical treatment of package switching systems anticipated the construction of computer networks which would function on a principle analogous to a post office rather than a telephone exchange: In a telephone system, there is a direct connection between the sender and receiver of information. But in a package switching system, there is no such connection — only the addresses of the sender and receiver on the package of information, which makes its way from node to node until it reaches its destination.

Further contributions to the concept of package switching systems and distributed communications networks were made by J.C.R. Licklider and W. Clark of MIT in 1962, and by Paul Baran of the RAND corporation in 1964. Licklider visualized what he called a "Galactic Network", a globally interconnected network of computers which would allow social interactions and interchange of data and software throughout the world. The distributed computer communication network proposed by Baran was motivated by the desire to have a communication system that could survive a nuclear war. The Cold War had also provoked the foundation (in 1957) of the Advanced Research Projects Agency (ARPA) by the U.S. government as a response to the successful Russian satellite "Sputnik".

In 1969, a 4-node network was tested by ARPA. It connected computers at the University of California divisions at Los Angeles and Santa Barbara with computers at the Stanford Research Institute and the University of Utah. Describing this event, Leonard Kleinrock said in an interview: "We set up a telephone connection between us and the guys at SRI. We typed the L and we asked on the phone 'Do you see the L?' 'Yes we see the L', came the response. We typed the 0 and we asked 'Do you see the 0?' 'Yes we see the O.' Then we typed the G and the system crashed". The ARPANET (with 40 nodes) performed much better in 1972 at the Washington Hilton Hotel where the participants at a Conference on Computer Communications were invited to test it.

Although the creators of ARPANET visualized it as being used for long-distance computations involving several computers, they soon discovered that social interactions over the Internet would become equally important if not more so. An electronic mail system was introduced in the early 1970's, and in 1976 Queen Elizabeth II of the United Kingdom became one of the increasing number of e-mail users.

In September, 1973, Robert F. Kahn and Vinton Cerf presented the basic ideas of the Internet at a meeting of the International Network Working Group at the University Sussex in Brighton, England. Among these principles was the rule that the networks to be connected should not be changed internally. Another rule was that if a packet did not arrive at its destination, it would be retransmitted from its original source. No information was to be retained by the gateways used to connect networks; and finally there was to be no global control of the Internet at the operations level.

Computer networks devoted to academic applications were introduced in the 1970's and 1980's, both in England, the United States and Japan. The Joint Academic Network (JANET) in the U.K. had its counterpart in the National Science Foundation's network (NSFNET) in America and Japan's JUNET (Japan Unix Network). Internet traffic is approximately doubling each year,[3] and it is about to overtake voice communication in

---

[3] In the period 1995–1996, the rate of increase was even faster — a doubling every four months.

the volume of information transferred.

In March, 2011, there were more than two billion Internet users in the world. In North America they amounted to 78.3 % of the total population, in Europe 58.3 % and worldwide, 30.2 %.

Another index that can give us an impression of the rate of growth of digital data generation and exchange is the "digital universe", which is defined to be the total volume of digital information that human information technology creates and duplicates in a year. In 2011, the digital universe reached 1.2 zettabytes, and it is projected to quadruple by 2015. A zettabyte is 10^{21} bytes, an almost unimaginable number, equivalent to the information contained in a thousand trillion books, enough books to make a pile that would stretch twenty billion kilometers.

Self-reinforcing information accumulation

Humans have been living on the earth for roughly two million years (more or less, depending on where one draws the line between our human and prehuman ancestors, Table 6.1). During almost all of this,time, our ancestors lived by hunting and food-gathering. They were not at all numerous, and did not stand out conspicuously from other animals. Then, suddenly, during the brief space of ten thousand years, our species exploded in numbers from a few million to seven billion (Figure 6.1), populating all parts of the earth, and even setting foot on the moon. This population explosion, which is still going on, has been the result of dramatic cultural changes. Genetically we are almost identical with our hunter-gatherer ancestors, who lived ten thousand years ago, but cultural evolution has changed our way of life beyond recognition.

Beginning with the development of speech, human cultural evolution began to accelerate. It started to move faster with the agricultural revolution, and faster still with the invention of writing and printing. Finally, modern science has accelerated the rate of social and cultural change to a completely unprecedented speed.

The growth of modern science is accelerating because knowledge feeds on itself. A new idea or a new development may lead to several other innovations, which can in turn start an avalanche of change. For example, the quantum theory of atomic structure led to the invention of transistors, which made high-speed digital computers possible. Computers have not only produced further developments in quantum theory; they have also revolutionized many other fields.

Table 7.1: Historical total world Internet traffic (after Cisco Visual
Networking Index Forecast). 1 terabyte =1,000,000,000,000 bytes

Year	Terabytes per month
1990	1
1991	2
1992	4
1993	10
1994	20
1995	170
1996	1,800
1997	5,000
1998	11,000
1999	26,000
2000	75,000
2001	175,000
2002	358,000
2003	681,000
2004	1,267,000
2005	2,055,000
2006	3,339,000
2007	5,219,000
2008	7,639,000
2009	10,676,000
2010	14,984,000

The self-reinforcing accumulation of knowledge — the information explosion — which characterizes modern human society is reflected not only
in an explosively-growing global population, but also in the number of scientific articles published, which doubles roughly every ten years. Another
example is Moore's law — the doubling of the information density of integrated circuits every two years. Yet another example is the explosive
growth of Internet traffic shown in Table 7.1.

The Internet itself is the culmination of a trend towards increasing societal information exchange — the formation of a collective human consciousness. This collective consciousness preserves the observations of millions of
eyes, the experiments of millions of hands, the thoughts of millions of brains;

and it does not die when the individual dies.

Suggestions for further reading

(1) H. Babbage, *Babbage's Calculating Engines: A Collection of Papers by Henry Prevost Babbage*, MIT Press, (1984).

(2) A.M. Turing, *The Enigma of Intelligence*, Burnett, London (1983).

(3) Ft. Penrose, *The Emperor's New Mind: Concerning Computers, Minds, and the Laws of Physics*, Oxford University Press, (1989).

(4) S. Wolfram, *A New Kind of Science*, Wolfram Media, Champaign IL, (2002).

(5) A.M. Turing, *On computable numbers, with an application to the Entscheidungsproblem*, Proc. Lond. Math. Soc. Ser 2, **42**, (1937). Reprinted in M. David Ed., *The Undecidable*, Raven Press, Hewlett N.Y., (1965).

(6) N. Metropolis, J. Howlett, and Gian-Carlo Rota (editors), *A History of Computing in the Twentieth Century*, Academic Press (1980).

(7) J. Shurkin, *Engines of the Mind: A History of Computers*, W.W. Norten, (1984).

(8) J. Palfreman and D. Swade, *The Dream Machine: Exploring the Computer Age*, BBC Press (UK), (1991).

(9) T.J. Watson, Jr. and P. Petre, *Father, Son, and Co.*, Bantam Books, New York, (1991).

(10) A. Hodges, Alan Turing: *The Enegma*, Simon and Schuster, (1983).

(11) H.H. Goldstein, *The Computer from Pascal to Von Neumann*, Princeton University Press, (1972).

(12) C.J. Bashe, L.R. Johnson, J.H. Palmer, and E.W. Pugh, *IBM's Early Computers*, Vol. 3 in the History of Computing Series, MIT Press, (1986).

(13) K.D. Fishman, *The Computer Establishment*, McGraw-Hill, (1982).

(14) S. Levy, *Hackers*, Doubleday, (1984).

(15) S. Franklin, *Artificial Minds*, MIT Press, (1997).

(16) P. Freiberger and M. Swaine, *Fire in the Valley: The Making of the Personal Computer*, Osborne/McGraw-Hill, (1984).

(17) R.X. Cringely, *Accidental Empires*, Addison-Wesley, (1992).

(18) R. Randell editor, *The Origins of Digital Computers, Selected Papers*, Springer-Verlag, New York (1973).

(19) H. Lukoff, *From Dits to Bits*, Robotics Press, (1979).

(20) D.E. Lundstrom, *A Few Good Men from Univac*, MIT Press, (1987).

(21) D. Rutland, *Why Computers Are Computers (The SWAC and the PC)*, Wren Publishers, (1995). .

(22) P.E. Ceruzzi, *Reckoners: The Prehistory of the Digital Computer, from Relays to the Stored Program Concept, 1935-1945*, Greenwood Press, Westport, (1983)

(23) S.G. Nash, *A History of Scientific Computing*, Addison-Wesley, Reading Mass., (1990).

(24) P.E. Ceruzzi, *Crossing the divide: Architectural issues and the emergence of stored programme computers*, 1935-1953, IEEE Annals of the History of Computing, **19**, 5-12, January-March (1997).

(25) P.E. Ceruzzi, *A History of Modern Computing*, MIT Press, Cambridge MA, (1998).

(26) K. Zuse, *Some remarks on the history of computing in Germany*, in *A History of Computing in the 20th Century*, N. Metropolis et al. editors, 611-627, Academic Press, New York, (1980).

(27) A.R. Mackintosh, *The First Electronic Computer*, Physics Today, March, (1987).

(28) S.H. Hollingdale and G.C. Tootil, *Electronic Computers*, Penguin Books Ltd. (1970).

(29) A. Hodges, *Alan Turing: The Enegma*, Simon and Schuster, New York, (1983).

(30) A. Turing, *On computable numbers with reference to the Entscheidungsproblem*, Journal of the London Mathematical Society, **II, 2. 42**, 230-265 (1937).

(31) J. von Neumann, *The Computer and the Brain*, Yale University Press, (1958).

(32) I.E. Sutherland, *Microelectronics and computer science*, Scientific American, 210-228, September (1977).

(33) W. Aspray, *John von Neumann and the Origins of Modern Computing*, M.I.T. Press, Cambridge MA, (1990, 2nd ed. 1992).

(34) W. Aspray, *The history of computing within the history of information technology*, History and Technology, **11**, 7-19 (1994).

(35) G.F. Luger, *Computation and Intelligence: Collected Readings*, MIT Press, (1995).

(36) Z.W. Pylyshyn, *Computation and Cognition: Towards a Foundation for Cognitive Science*, MIT Press, (1986).

(37) D.E. Shasha and C. Lazere, *Out of Their Minds: The Creators of Computer Science*, Copernicus, New York, (1995).

(38) W. Aspray, *An annotated bibliography of secondary sources on the*

history of software, Annals of the History of Computing **9**, 291-243 (1988).

(39) R. Kurzweil, *The Age of Intelligent Machines*, MIT Press, (1992).

(40) S.L. Garfinkel and H. Abelson, eds., *Architects of the Information Society: Thirty-Five Years of the Laboratory for Computer Sciences at MIT*, MIT Press, (1999).

(41) J. Haugeland, *Artificial Intelligence: The Very Idea*, MIT Press, (1989).

(42) M.A. Boden, *Artificial Intelligence in Psychology: Interdisciplinary Essays*, MIT Press, (1989).

(43) J.W. Cortada, *A Bibliographic Guide to the History of Computer Applications, 1950-1990*, Greenwood Press, Westport Conn., (1996).

(44) M. Campbell-Kelly and W. Aspry, *Computer: A History of the Information Machine*, Basic Books, New York, (1996).

(45) B.I. Blum and K. Duncan, editors, *A History of Medical Informatics*, ACM Press, New York, (1990).

(46) J.-C. Guedon, La Planete Cyber, *Internet et Cyberspace*, Gallimard, (1996).

(47) S. Augarten, *Bit by Bit: An Illustrated History of Computers*, Unwin, London, (1985).

(48) N. Wiener, *Cybernetics; or Control and Communication in the Animal and the Machine*, The Technology Press, John Wiley and Sons, New York, (1948).

(49) W.R. Ashby, *An Introduction to Cybernetics*, Chapman and Hall, London, (1956).

(50) M.A. Arbib, *A partial survey of cybernetics in eastern Europe and the Soviet Union*, Behavioral Sci., **11**, 193-216, (1966).

(51) A. Rosenblueth, N. Weiner and J. Bigelow, *Behavior, purpose and teleology*, Phil. Soc. **10** (1), 18-24 (1943).

(52) N. Weiner and A. Rosenblueth, *Conduction of impulses in cardiac muscle*, Arch. Inst. Cardiol. Mex., **16**, 205-265 (1946).

(53) H. von Foerster, editor, *Cybernetics - circular, causal and feed-back mechanisms in biological and social systems. Transactions of sixthtenth conferences*, Josiah J. Macy Jr. Foundation, New York, (1950-1954).

(54) W.S. McCulloch and W. Pitts, *A logical calculus of ideas immanent in nervous activity*, Bull. Math. Biophys., **5**, 115-133 (1943).

(55) W.S. McCulloch, *An Account of the First Three Conferences on Teleological Mechanisms*, Josiah Macy Jr. Foundation, (1947).

(56) G.A. Miller, *Languages and Communication*, McGraw-Hill, New York, (1951).

(57) G.A. Miller, *Statistical behavioristics and sequences of responses*, Psychol. Rev. **56**, 6 (1949).

(58) G. Bateson, *Bali — the value system of a steady state*, in M. Fortes, editor, *Social Structure Studies Presented to A.R. Radcliffe-Brown*, Clarendon Press, Oxford, (1949).

(59) G. Bateson, *Communication, the Social Matrix of Psychiatry*, Norton, (1951).

(60) G. Bateson, *Steps to an Ecology of Mind*, Chandler, San Francisco, (1972).

(61) G. Bateson, *Communication et Societé*, Seuil, Paris, (1988).

(62) S. Heims, *Gregory Bateson and the mathematicians: From interdisciplinary interactions to societal functions*, J. History Behavioral Sci., **13**, 141-159 (1977).

(63) S. Heims, *John von Neumann and Norbert Wiener. From Mathematics to the Technology of Life and Death*, MIT Press, Cambridge MA, (1980).

(64) S. Heims, *The Cybernetics Group*, MIT Press, Cambridge MA, (1991).

(65) G. van de Vijver, *New Perspectives on Cybernetics (Self-Organization, Autonomy and Connectionism)*, Kluwer, Dordrecht, (1992).

(66) A. Bavelas, *A mathematical model for group structures*, Appl. Anthrop. **7 (3)**, 16 (1948).

(67) P. de Latil, *La Pensée Artificielle — Introduction a la Cybernetique*, Gallimard, Paris, (1953).

(68) L.K. Frank, G.E. Hutchinson, W.K. Livingston, W.S. McCulloch and N. Wiener, *Teleological Mechanisms*, Ann. N.Y. Acad. Sci. **50**, 187-277 (1948).

(69) H. von Foerster, *Quantum theory of memory*, in H. von Foerster, editor, *Cybernetics — circular, causal and feed-back mechanisms in biological and social systems*. Transactions of the sixth conferences, Josiah J. Macy Jr. Foundation, New York, (1950).

(70) H. von Foerster, *Observing Systems*, Intersystems Publications, California, (1984).

(71) H. von Foerster, *Understanding Understanding: Essays on Cybernetics and Cognition*, Springer, New York, (2002).

(72) M. Newborn, *Kasparov vs. Deep Blue: Computer Chess Comes of*

age, Springer Verlag, (1996).

(73) K.M. Colby, *Artificial Paranoia: A Computer Simulation of the Paranoid Process*, Pergamon Press, New York, (1975).

(74) J.Z. Young, *Discrimination and learning in the octopus*, in H. von Foerster, editor, *Cybernetics — circular, causal and feed-back mechanisms in biological and social systems*. Transactions of the ninth conference, Josiah J. Macy Jr. Foundation, New York, (1953).

(75) M.J. Apter and L. Wolpert, *Cybernetics and development*. I. Information theory, J. Theor. Biol. **8**, 244-257 (1965).

(76) H. Atlan, *L'Organization Biologique et la Theorie de I'Information*, Hermann, Paris, (1972).

(77) H. Atlan, *On a formal definition of organization*, J. Theor. Biol. 45, 295-304 (1974).

(78) H. Atlan, *Organization du vivant, information et auto-organization*, in Volume Symposium 1986 de l'Encylopediea Universalis, pp. 355-361, Paris, (1986).

(79) E.R. Kandel, *Nerve cells and behavior*, Scientific American, **223**, 57-70, July, (1970).

(80) E.R. Kandel, *Small systems of neurons*, Scientific American, **241 no.3**, 66-76 (1979).

(81) A.K. Katchalsky et al., *Dynamic patterns of brain cell assemblies*, Neurosciences Res. Prog. Bull., **12 no.1**, (1974).

(82) G.E. Moore, *Cramming more components onto integrated circuits*, Electronics, April 19, (1965).

(83) P. Gelsinger, P. Gargini, G. Parker and A. Yu, *Microprocessors circa 2000*, IEEE Spectrum, October, (1989).

(84) P. Baron, *On distributed communications networks*, IEEE Trans. Comm. Systems, March (1964).

(85) V.G. Cerf and R.E. Khan, *A protocol for packet network intercommunication*, Trans. Comm. Tech. **COM-22, V5**, 627-641, May (1974).

(86) L. Kleinrock, *Communication Nets: Stochastic Message Flow and Delay*, McGraw-Hill, New York, (1964).

(87) L. Kleinrock, *Queueing Systems: Vol. II, Computer Applications*, Wiley, New York, (1976).

(88) R. Kahn, editor, *Special Issue on Packet Communication Networks*, Proc. IEEE, **66**, November, (1978).

(89) L.G. Roberts, *The evolution of packet switching*, Proc. of the IEEE **66**, 1307-13, (1978).

(90) J. Abbate, *The electrical century: Inventing the web*, Proc. IEEE **87**, November, (1999).

(91) J. Abbate, *Inventing the Internet*, MIT Press, Cambridge MA, (1999).

(92) J.C. McDonald, editor, *Fundamentals of Digital Switching, 2nd Edition*, Plenum, New York, (1990).

(93) B. Metcalfe, *Packet Communication*, Peer-to-Peer Communication, San Jose Calif, (1996).

(94) T. Berners-Lee, *The Original Design and Ultimate Destiny of the World Wide Web by its Inventor*, Harper San Francisco, (1999).

(95) J. Clark, *Netscape Time: The Making of the Billion-Dollar Start-Up That Took On Microsoft*, St. Martin's Press, New York, (1999).

(96) J. Wallace, *Overdrive: Bill Gates and the Race to Control Cyberspace*, Wiley, New York, (1997).

(97) P. Cunningham and F. Froschl, *The Electronic Business Revolution*, Springer Verlag, New York, (1999).

(98) J.L. McKenny, *Waves of Change: Business Evolution Through Information Technology*, Harvard Business School Press, (1995).

(99) M.A. Cosumano, *Competing on Internet Time: Lessons From Netscape and Its Battle with Microsoft*, Free Press, New York, (1998).

(100) F.J. Dyson, *The Sun, the Genome and the Internet: Tools of Scientific Revolutions*, Oxford University Press, (1999).

(101) L. Bruno, *Fiber Optimism: Nortel, Lucent and Cisco are battling to win the high-stakes fiber-optics game*, Red Herring, June (2000).

(102) N. Cochrane, *We're insatiable: Now it's 20 million million bytes a day*, Melbourne Age, January 15, (2001).

(103) K.G. Coffman and A.N. Odlyzko, The size and growth rate of the Internet, First Monday, October, (1998).

(104) C.A. Eldering, M.L. Sylla, and J.A. Eisenach, *Is there a Moore's law for bandwidth?*, IEEE Comm. Mag., 2-7, October, (1999).

(105) G. Gilder, *Fiber keeps its promise: Get ready, bandwidth will triple each year for the next 25 years*, Forbes, April 7, (1997).

(106) M. Weisner, *The computer for the 21st century*, Scientific American, September, (1991).

(107) R. Wright, *Three Scientists and Their Gods*, Time Books, (1988).

(108) S. Nora and A. Mine, *The Computerization of Society*, MIT Press, (1981).

(109) T. Forester, *Computers in the Human Context: Information Theory, Productivity, and People*, MIT Press, (1989).

(110) W. Shockley, *Electrons and Holes in Semiconductors with Applications to Transistor Electronics*, Bell Laboratories Series, (1950).

Chapter 8

BIO-INFORMATION
TECHNOLOGY

The merging of information technology and biotechnology

Information technology and biology are today the two most rapidly developing fields of science. Interestingly, these two fields seem to be merging, each gaining inspiration and help from the other. For example, computer scientists designing both hardware and software are gaining inspiration from physiological studies of the mechanism of the brain; and conversely, neurophysiologists are aided by insights from the field of artificial intelligence. Designers of integrated circuits wish to prolong the period of validity of Moore's law; but they are rapidly approaching physical barriers which will set limits to the miniaturization of conventional transistors and integrated circuits. They gain inspiration from biology, where the language of molecular complementarity and the principle of autoassembly seem to offer hope that molecular switches and self-assembled integrated circuits may one day be constructed.

Geneticists, molecular biologists, biochemists and crystallographers have now obtained so much information about the amino acid sequences and structures of proteins and about the nucleotide sequences in genomes that the full power of modern information technology is needed to store and to analyze this information. Computer scientists, for their part, turn to evolutionary genetics for new and radical methods of developing both software and hardware — genetic algorithms and simulated evolution.

Self-assembly of supramolecular structures — Nanoscience

In previous chapters, we saw that the language of molecular complementarity (the "lock and key" fitting discovered by Paul Ehrlich) is the chief mechanism by which information is stored and transferred in biological systems. Biological molecules have physical shapes and patterns of excess charge[1] which are recognized by complementary molecules because they fit together, just as a key fits the shape of a lock. Examples of biological "lock and key" fitting are the fit between the substrate of an enzyme and the enzyme's active site, the recognition of an antigen by its specific antibody, the specificity of base pairs in DNA and RNA, and the autoassembly of structures such as viruses and subcellular organelles.

One of the best studied examples of autoassembly through the mechanism of molecular complementarity is the tobacco mosaic virus. The assembled virus has a cylindrical form about 300 nm long (1 nm = 1 nanometer = 10^{-9} meters = 10 Ångstroms), with a width of 18 nm. The cylindrically shaped virus is formed from about 2000 identical protein molecules. These form a package around an RNA molecule with a length of approximately 6400 nucleotides. The tobacco mosaic virus can be decomposed into its constituent molecules in vitro, and the protein and RNA can be separated and put into separate bottles, as was discussed in Chapter 4.

If, at a later time, one mixes the protein and RNA molecules together in solution, they spontaneously assemble themselves into new infective tobacco mosaic virus particles. The mechanism for this spontaneous autoassembly is a random motion of the molecules through the solvent until they approach each other in such a way that a fit is formed. When two molecules fit closely together, with their physical contours matching, and with complementary patterns of excess charge also matching, the Gibbs free energy of the total system is minimized. Thus the self-assembly of matching components proceeds spontaneously, just as every other chemical reaction proceeds spontaneously when the difference in Gibbs free energy between the products and reactants is negative. The process of autoassembly is analogous to crystallization, except that the structure formed is more complex than an ordinary crystal.

A second very well-studied example of biological autoassembly is the spontaneous formation of bilayer membranes when phospholipid molecules are shaken together in water. Each phospholipid molecule has a small polar (hydrophilic) head, and a long nonpolar (hydrophobic) tail. The polar head is hydrophilic — water-loving — because it has large excess charges with which water can form hydrogen bonds. By contrast, the non-polar tail of a phospholipid molecule has no appreciable excess charges. The tail is hydrophobic — it hates water — because to fit into the water structure it

---

[1] They also have patterns of polarizable groups and reactive groups, and these patterns can also play a role in recognition.

has to break many hydrogen bonds to make a hole for itself, but it cannot pay for these broken bonds by forming new hydrogen bonds with water.

There is a special configuration of the system of water and phospholipid molecules which has a very low Gibbs free energy — the lipid bilayer. In this configuration, all the hydrophilic polar heads are in contact with water, while the hydrophobic nonpolar tails are in the interior of the double membrane, away from the water, and in close contact with each other, thus maximizing their mutual Van der Waals attractions. (The basic structure of biological membranes is the lipid bilayer just described, but there are also other components, such as membrane-bound proteins, caveolae, and ion pores.)

The mechanism of self-organization of supramolecular structures is one of the most important universal mechanisms of biology. Chemical reactions take place spontaneously when the change in Gibbs free energy produced by the reaction is negative, i.e., chemical reactions take place in such a direction that the entropy of the universe increases. When spontaneous chemical reactions take place, the universe moves from a less probable configuration to a more probable one. The same principle controls the motion of larger systems, where molecules arrange themselves spontaneously to form supramolecular structures. Self-assembling collections of molecules move in such a way as to minimize their Gibbs free energy, thus maximizing the entropy of the universe.

Biological structures of all kinds are formed spontaneously from their components because assembly information is written onto their joining surfaces in the form of complementary surface contours and complementary patterns of excess charge[2]. Matching pieces fit together, and the Gibbs free energy of the system is minimized. Virtually every structure observed in biology is formed in this way — by a process analogous to crystallization, except that biological structures can be far more complex than ordinary crystals.

Researchers in microelectronics, inspired by the self-assembly of biological structures, dream of using the same principles to generate self-organizing integrated circuits with features so small as to approach molecular dimensions. As we mentioned in Chapter 7, the speed of a computing operation is limited by the time that it takes an electrical signal (moving at approximately the speed of light) to traverse a processing unit. The desire to produce ever greater computation speeds as well as ever greater memory

---

[2] Patterns of reactive or polarizable groups also play a role.

densities, motivates the computer industry's drive towards ultraminiaturization.

Currently the fineness of detail in integrated circuits is limited by diffraction effects caused by the finite wavelength of the light used to project an image of the circuit onto a layer of photoresist covering the chip where the circuit is being built up. For this reason, there is now very active research on photolithography using light sources with extremely short wavelengths, in the deep ultraviolet, or even X-ray sources, synchrotron radiation, or electron beams. The aim of this research is to produce integrated circuits whose feature size is in the nanometer range — smaller than 100 nm. In addition to these efforts to create nanocircuits by "top down" methods, intensive research is also being conducted on "bottom up" synthesis, using principles inspired by biological self-assembly. The hope to make use of "the spontaneous association of molecules, under equilibrium conditions, into stable, structurally well-defined aggregates, joined by non-covalent bonds"[3].

The Nobel Laureate Belgian chemist J.-M. Lehn pioneered the field of supramolecular chemistry by showing that it is possible to build nanoscale structures of his own design. Lehn and his coworkers at the University of Strasbourg used positively-charged metal ions as a kind of glue to join larger structural units at points where the large units exhibited excess negative charges. Lehn predicts that the supramolecular chemistry of the future will follow the same principles of self-organization which underlie the growth of biological structures, but with a greatly expanded repertoire, making use of elements (such as silicon) that are not common in carbon-based biological systems.

Other workers in nanotechnology have concentrated on the self-assembly of two-dimensional structures at water-air interfaces. For example, Thomas Bjørnholm, working at the University of Copenhagen, has shown that a nanoscale wire can be assembled spontaneously at a water-air interface, using metal atoms complexed with DNA and a DNA template. The use of a two-dimensional template to reproduce a nanostructure can be thought of as "microprinting". One can also think of self-assembly at surfaces as the two-dimensional version of the one-dimensional copying process by which a new DNA or RNA strand assembles itself spontaneously, guided by the complementary strand.

In 1981, Gerd Binning and Heinrich Rohrer of IBM's Research Center in Switzerland announced their invention of the scanning tunneling micro-

---

[3] G.M. Whiteside et al., Science, **254**, 1312-1314, (1991).

scope. The new microscope's resolution was so great that single atoms could be observed. The scanning tunneling microscope consists of a supersharp conducting tip, which is brought near enough to a surface so that quantum mechanical tunneling of electrons can take place between tip and surface when a small voltage is applied. The distance between the supersharp tip and the surface is controlled by means of a piezoelectric crystal. As the tip is moved along the surface, its distance from the surface (and hence the tunneling current) is kept constant by applying a voltage to the piezoelectric crystal, and this voltage as a function of position gives an image of the surface.

Variations on the scanning tunneling microscope allow single atoms to be deposited or manipulated on a surface. Thus there is a hope that nanoscale circuit templates can be constructed by direct manipulation of atoms and molecules, and that the circuits can afterwards be reproduced using autoassembly mechanisms.

The scanning tunneling microscope makes use of a quantum mechanical effect: Electrons exhibit wavelike properties, and can tunnel small distances into regions of negative kinetic energy — regions which would be forbidden to them by classical mechanics. In general it is true that for circuit elements with feature sizes in the nanometer range, quantum effects become important. For conventional integrated circuits, the quantum effects which are associated with this size-range would be a nuisance, but workers in nanotechnology hope to design integrated circuits which specifically make use of these quantum effects.

Molecular switches; bacteriorhodopsin

The purple, salt-loving archaebacterium Halobacterium halobium (recently renamed Halobacterium salinarum) possesses one of the simplest structures that is able to perform photosynthesis. The purple membrane subtraction of this bacterium's cytoplasmic membrane contains only two kinds of molecules — lipids and bacteriorhodopsin. Nevertheless, this simple structure is able to trap the energy of a photon from the sun and to convert it into chemical energy.

The remarkable purple membrane of Halobacterium has been studied in detail by Walter Stoeckenius, D. Osterhelt[4], Lajos Keszthelyi and others.

---

[4] D. Osterhelt and Walter Stoeckenius, Nature New Biol. **233**, 149-152 (1971); D. Osterhelt et al., Quart. Rev. Biophys. **24**, 425-478 (1991); W. Stoeckenius and R. Bogomolni, Ann. Rev. Biochem. **52**, 587-616 (1982).

It can be decomposed into its constituent molecules. The lipids from the membrane and the bacteriorhodopsin can be separated from each other and put into different bottles. At a later time, the two bottles can be taken from the laboratory shelf, and their contents can be shaken together in water. The result is the spontaneous formation of tiny vesicles of purple membrane.

In the self-organized two-component vesicles, the membrane-bound protein bacteriorhodopsin is always correctly oriented, just as it would be in the purple membrane of a living Halobacterium. When the vesicles are illuminated, bacteriorhodopsin absorbs H^+ ions from the water on the inside, and releases them outside.

Bacteriorhodopsin consists of a chain of 224 amino acids, linked to the retinal chromophore. The amino acids are arranged in 7 helical segments, each of which spans the purple membrane, and these are joined on the membrane surface by short nonhelical segments of the chain. The chromophore is in the middle of the membrane, surrounded by α-helical segments. When the chromophore is illuminated, its color is temporarily bleached, and it undergoes a cis-trans isomerization which disrupts the hydrogen-bonding network of the protein. The result is that a proton is released on the outside of the membrane. Later, a proton is absorbed from the water in the interior of the membrane vesicle, the hydrogen-bonding system of the protein is reestablished, and both the protein and the chromophore return to their original conformations. In this way, bacteriorhodopsin functions as a proton pump. It uses the energy of photons to transport H^+ ions across the membrane, from the inside to the outside, against the electrochemical gradient. In the living Halobacterium, this H^+ concentration difference would be used to drive the synthesis of the high-energy phosphate bond of adenosine triphosphate (ATP), the inward passage of H^+ through other parts of the cytoplasmic membrane being coupled to the reaction $ADP + P_i \rightarrow ATP$ by membrane-bound reversible ATPase.

Bacteriorhodopsin is interesting as a component of one of the simplest known photosynthetic systems, and because of its possible relationship to the evolution of the eye (as was discussed in Chapter 3). In addition, researchers like Lajos Keszthelyi at the Institute of Biophysics of the Hungarian Academy of Sciences in Szeged are excited about the possible use of bacteriorhodopsin in optical computer memories[5]. Arrays of oriented and partially dehydrated bacteriorhodopsin molecules in a plastic matrix can be

---

[5] A. Der and L. Keszthelyi, editors, Bioelectronic Applications of Photochromic Pigments, IOS Press, Amsterdam, Netherlands, (2001).

used to construct both 2-dimensional and 3-dimensional optical memories using the reversible color changes of the molecule. J. Chen and coworkers[6] have recently constructed a prototype 3-dimensional optical memory by orienting the proteins and afterwards polymerizing the solvent into a solid polyacrylamide matrix. Bacteriorhodopsin has extraordinary stability, and can tolerate as many as a million optical switching operations without damage.

Neural networks, biological and artificial

In 1943, W. McCulloch and W. Pitts published a paper entitled *A Logical Calculus of the Ideas Immanent in Nervous Activity*. In this pioneering paper, they proposed the idea of a Threshold Logic Unit (TLU), which they visualized not only as a model of the way in which neurons function in the brain but also as a possible subunit for artificial systems which might be constructed to perform learning and pattern-recognition tasks. Problems involving learning, generalization, pattern recognition and noisy data are easily handled by the brains of humans and animals, but computers of the conventional von Neumann type find such tasks especially difficult.

Conventional computers consist of a memory and one or more central processing units (CPUs). Data and instructions are repeatedly transferred from the memory to the CPUs, where the data is processed and returned to the memory. The repeated performance of many such cycles requires a long and detailed program, as well as high-quality data. Thus conventional computers, despite their great speed and power, lack the robustness, intuition, learning powers and powers of generalization which characterize biological neural networks. In the 1950's, following the suggestions of McCulloch and Pitts, and inspired by the growing knowledge of brain structure and function which was being gathered by histologists and neurophysiologists, computer scientists began to construct artificial neural networks - massively parallel arrays of TLU's.

The analogy between a TLU and a neuron can be seen by comparing Figure 5.2, which shows a neuron, with Figure 8.1, which shows a TLU. As we saw in Chapter 5, a neuron is a specialized cell consisting of a cell body (*soma*) from which an extremely long, tubelike fiber called an *axon* grows. The axon is analogous to the output channel of a TLU. From the soma, a number of slightly shorter, rootlike extensions called *dendrites* also grow. The dendrites are analogous to the input channels of a TLU.

---

[6] J. Chen et al., Biosystems **35**, 145-151 (1995).

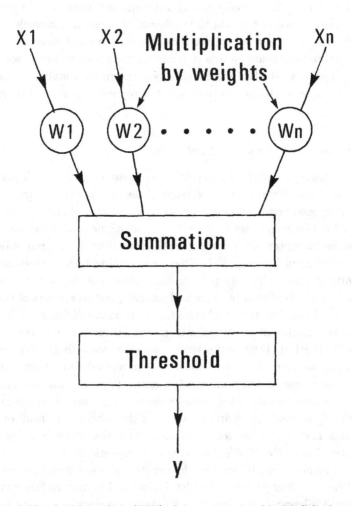

Fig. 8.1   A Threshold Logic Unit (TLU) of the type proposed by McCulloch and Pitts.

In a biological neural network, branches from the axon of a neuron are connected to the dendrites of many other neurons; and at the points of connection there are small, knoblike structures called synapses. As was discussed in Chapter 5, the "firing" of a neuron sends a wave of depolarization out along its axon. When the pulselike electrical and chemical disturbance associated with the wave of depolarization (the action potential) reaches a synapse, where the axon is connected with another neuron, transmitter molecules are released into the post-synaptic cleft. The neurotransmitter

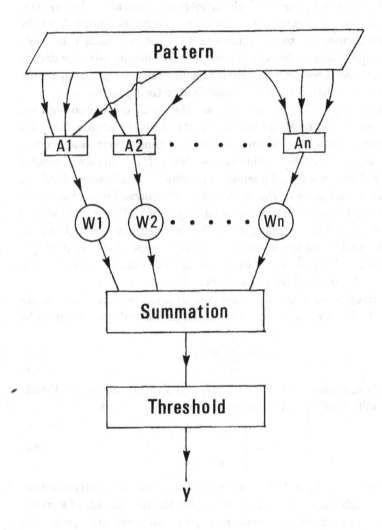

Fig. 8.2 A perceptron, introduced by Rosenblatt in 1962. The perceptron is similar to a TLU, but its input is preprocessed by a set of association units (A-units). The A-units are not trained, but are assigned a fixed Boolean functionality.

molecules travel across the post-synaptic cleft to receptors on a dendrite of the next neuron in the net, where they are bound to receptors. There are many kinds of neurotransmitter molecules, some of which tend to make the firing of the next neuron more probable, and others which tend to inhibit its firing. When the neurotransmitter molecules are bound to the receptors, they cause a change in the dendritic membrane potential, either increasing or decreasing its polarization. The post-synaptic potentials from the dendrites are propagated to the soma; and if their sum exceeds a threshold value, the neuron fires. The subtlety of biological neural networks derives from the fact that there are many kinds of neurotransmitters and synapses, and from the fact that synapses are modified by their past history.

Turning to Figure 8.1, we can compare the biological neuron with the Threshold Logic Unit of McCulloch and Pitts. Like the neuron, the TLU has many input channels. To each of the N channels there is assigned a weight, $w_1, w_2, ..., w_N$. The weights can be changed; and the set of weights gives the TLU its memory and learning capabilities. Modification of weights in the TLU is analogous to the modification of synapses in a neuron, depending on their history. In the most simple type of TLU, the input signals are either 0 or 1. These signals, multiplied by their appropriate weights, are summed, and if the sum exceeds a threshold value, θ the TLU "fires", i.e., a pulse of voltage is transmitted through the output channel to the next TLU in the artificial neural network.

Let us imagine that the input signals, $x_1, x_2, ..., x_N$ can take on the values 0 or 1. The weighted sum of the input signals will then be given by

$$a = \sum_{j=1}^{N} w_j x_j \qquad (8.1)$$

The quantity a, is called the *activation*. If the activation exceeds the threshold 9, the unit "fires", i.e., it produces an output y given by

$$y = \begin{cases} 1 \text{ if } a \geq \theta \\ \\ 0 \text{ if } a < \theta \end{cases} \qquad (8.2)$$

The decisions taken by a TLU can be given a geometrical interpretation: The input signals can be thought of as forming the components of a vector, $x = x_1, x_2, ..., x_N$, in an N-dimensional space called pattern space. The weights also form a vector, $w = w_1, w_2, ..., w_N$, in the same space. If we write an equation setting the scalar product of these two vectors equal to some constant,

$$\mathbf{w} \cdot \mathbf{x} \equiv \sum_{j=1}^{N} w_j x_j = \theta \qquad (8.3)$$

then this equation defines a hyperplane in pattern space, called the *decision hyperplane*. The decision hyperplane divides pattern space into two parts:

(1) input pulse patterns which will produce firing of the TLU, and (2) patterns which will not cause firing.

The position and orientation of the decision hyperplane can be changed by altering the weight vector w and/or the threshold θ. Therefore it is convenient to put the threshold and the weights on the same footing by introducing an augmented weight vector,

$$\mathbf{W} = w_1, w_2, ..., w_N, \theta \tag{8.4}$$

and an augmented input pattern vector,

$$\mathbf{X} = x_1, x_2, ..., x_N, -1 \tag{8.5}$$

In the $N+1$-dimensional augmented pattern space, the decision hyperplane now passes through the origin, and equation (8.3) can be rewritten in the form

$$\mathbf{W} \cdot \mathbf{X} \equiv \sum_{j=1}^{N+1} W_j X_j = 0 \tag{8.6}$$

Those input patterns for which the scalar product $\mathbf{W} \cdot \mathbf{X}$ is positive or zero will cause the unit to fire, but if the scalar product is negative, there will be no response.

If we wish to "teach" a TLU to fire when presented with a particular pattern vector $\mathbf{X}$, we can evaluate its scalar product with the current augmented weight vector $\mathbf{W}$. If this scalar product is negative, the TLU will not fire, and therefore we know that the weight vector needs to be changed. If we replace the weight vector by

$$\mathbf{W}' = \mathbf{W} + \gamma \mathbf{X} \tag{8.7}$$

where γ is a small positive number, then the new augmented weight vector $\mathbf{W}'$ will point in a direction more nearly the same as the direction of $\mathbf{X}$. This change will be a small step in the direction of making the scalar product positive, i.e., a small step in the right direction.

Why not take a large step instead of a small one? A small step is best because there may be a whole class of input patterns to which we would like the TLU to respond by firing. If we make a large change in weights to help a particular input pattern, it may undo previous learning with respect to other patterns.

It is also possible to teach a TLU to remain silent when presented with a particular input pattern vector. To do so, we evaluate the augmented scalar product $\mathbf{W} \cdot \mathbf{X}$ as before, but now, when we desire silence rather than firing, we wish the scalar product to be negative, and if it is positive, we know that the weight vector must be changed. In changing the weight vector, we can again make use of equation (8.7), but now γ must be a small negative number rather than a small positive one.

Two sets of input patterns, A and B, are said to be linearly separable if they can be separated by some decision hyperplane in pattern space. Now suppose that the four sets, A, B, C, and D, can be separated by two decision hyperplanes. We can then construct a two-layer network which will identify the class of an input signal belonging to any one of the sets.

The first layer consists of two TLU's. The first TLU in this layer is taught to fire if the input pattern belongs to A or B, and to be silent if the input belongs to C or D. The second TLU is taught to fire if the input pattern belongs to A or D, and to be silent if it belongs to B or C. The second layer of the network consists of four output units which are not taught, but which are assigned a fixed Boolean functionality. The first output unit fires if the signals from the first layer are given by the vector $\mathbf{y} = \{0, 0\}$ (class A); the second fires if $\mathbf{y} = \{0, 1\}$ (class B), the third if $\mathbf{y} = \{1, 0\}$ (class C), and the fourth if $\mathbf{y} = \{1, 1\}$ (class D). Thus the simple two-layer network functions as a *classifier*. The output units in the second layer are analogous to the "grandmother's face cells" whose existence in the visual cortex is postulated by neurophysiologists. These cells will fire if and only if the retina is stimulated with a particular class of patterns.

This very brief glance at artificial neural networks does not do justice to the high degree of sophistication which network architecture and training algorithms have achieved during the last two decades. However, the suggestions for further reading at the end of this chapter may help to give the reader an impression of the wide range of problems to which these networks are now being applied.

Besides being useful for computations requiring pattern recognition, learning, generalization, intuition, and robustness in the face of noisy data, artificial neural networks are important because of the light which they throw on the mechanism of brain function. For example, one can compare the classifier network with the discoveries of Kuffler, Hubel and Wessel concerning pattern abstraction in the mammalian retina and visual cortex (Chapter 5).

Genetic algorithms

Genetic algorithms represent a second approach to machine learning and to computational problems involving optimization. Like neural network computation, this alternative approach has been inspired by biology, and it has also been inspired by the Darwinian concept of natural selection. In a genetic algorithm, the hardware is that of a conventional computer; but the software creates a population and allows it to evolve in a manner closely analogous to biological evolution.

One of the most important pioneers of genetic algorithms was John Henry Holland (1929–). After attending MIT, where he was influenced by Norbert Wiener, Holland worked for IBM, helping to develop the 701. He then continued his studies at the University of Michigan, obtaining the first Ph.D. in computer science ever granted in America. Between 1962 and 1965, Holland taught a graduate course at Michigan called "Theory of Adaptive Systems". His pioneering course became almost a cult, and together with his enthusiastic students he applied the genetic algorithm approach to a great variety of computational problems. One of Holland's students, David Goldberg, even applied a genetic algorithm program to the problem of allocating natural gas resources.

The programs developed by Holland and his students were modelled after the natural biological processes of reproduction, mutation, selection and evolution. In biology, the information passed between generations is contained in chromosomes — long strands of DNA where the genetic message is written in a four-letter language, the letters being adenine, thymine, guanine and cytosine. Analogously, in a genetic algorithm, the information is coded in a long string, but instead of a four-letter language, the code is binary: The chromosome-analogue is a long string of 0's and 1's, i.e., a long binary string. One starts with a population that has sufficient diversity so that natural selection can act.

The genotypes are then translated into phenotypes. In other words, the information contained in the long binary string (analogous to the genotype of each individual) corresponds to an entity, the phenotype, whose fitness for survival can be evaluated. The mapping from genotype to phenotype must be such that very small changes in the binary string will not produce radically different phenotypes. From the initial population, the most promising individuals are selected to be the parents of the next generation, and of these, the fittest are allowed produce the largest number of offspring. Before reproduction takes place, however, random mutations and chromosome crossing can occur. For example, in chromosome crossing, the chromosomes of two individuals are broken after the nth binary digit, and two new chromosomes are formed, one with the head of the first old chromosome and the tail of the second, and another with the head of the second and the tail of the first. This process is analogous to the biological crossings which allowed Thomas Hunt Morgan and his "fly squad" to map the positions of genes on the chromosomes of fruit flies, while the mutations are analogous to those studied by Hugo de Vries and Hermann J. Muller.

After the new generation has been produced, the genetic algorithm advances the time parameter by a step, and the whole process is repeated: The phenotypes of the new generation are evaluated and the fittest selected to be parents of the next generation; mutation and crossings occur; and then fitness-proportional reproduction. Like neural networks, genetic algorithms are the subject of intensive research, and evolutionary computation is a rapidly growing field.

Evolutionary methods have been applied not only to software, but also to hardware. Some of the circuits designed in this way defy analysis using conventional techniques — and yet they work astonishingly well.

Artificial life

As Aristotle pointed out, it is difficult to define the precise border between life and nonlife. It is equally difficult to give a precise definition of artificial life. Of course the term means "life produced by humans rather than by nature", but what is life? Is self-replication the only criterion? The phrase "produced by humans" also presents difficulties. Humans have played a role in creating domestic species of animals and plants. Can cows, dogs, and high-yield wheat varieties be called "artificial life" ? In one sense, they can. These species and varieties certainly would not have existed without human intervention.

We come nearer to what most people might call "artificial life" when we take parts of existing organisms and recombine them in novel ways, using the techniques of biotechnology. For example, Steen Willadsen[7], working at the Animal Research Station, Cambridge England, was able to construct chimeras by operating under a microscope on embryos at the eight-cell stage. The zona pelucida is a transparent shell that surrounds the cells of

---

[7] Willadsen is famous for having made the first verified and reproducible clone of a mammal. In 1984 he made two genetically identical lambs from early sheep embryo cells.

the embryo. Willadsen was able to cut open the zona pelucida, to remove the cells inside, and to insert a cell from a sheep embryo together with one from a goat embryo. The chimeras which he made in this way were able to grow to be adults, and when examined, their cells proved to be a mosaic, some cells carrying the sheep genome while others carried the genome of a goat. By the way, Willadsen did not create his chimeras in order to produce better animals for agriculture. He was interested in the scientifically exciting problem of morphogenesis: How is the information of the genome translated into the morphology of the growing embryo?

Human genes are now routinely introduced into embryos of farm animals, such as pigs or sheep. The genes are introduced into regulatory sequences which cause expression in mammary tissues, and the adult animals produce milk containing human proteins. Many medically valuable proteins are made in this way. Examples include human blood-clotting factors, interleukin-2 (a protein which stimulates T-lymphocytes), collagen and fibrinogen (used to treat burns), human fertility hormones, human hemoglobin, and human serum albumin.

Transgenic plants and animals in which the genes of two or more species are inherited in a stable Mendelian way have become commonplace in modern laboratory environments, and, for better or for worse, they are also becoming increasingly common in the external global environment. These new species might, with some justification, be called "artificial life".

In discussing the origin of life in Chapter 3, we mentioned that a long period of molecular evolution probably preceded the evolution of cells. In the early 1970's, S. Spiegelman performed a series of experiments in which he demonstrated that artificial molecular evolution can be made to take place in vitro. Spiegelman prepared a large number of test tubes in which RNA replication could take place. The aqueous solution in each of the test tubes consisted of RNA replicase, ATP, UTP (uracil triphosphate), GTP (guanine triphosphate), CTP (cytosine triphosphate) and buffer. He then introduced RNA from a bacteriophage into the first test tube. After a predetermined interval of time, during which replication took place, Spiegelman transferred a drop of solution from the first test tube to a new tube, uncontaminated with RNA. Once again, replication began and after an interval a drop was transferred to a third test tube. Spiegelman repeated this procedure several hundred times, and at the end he was able to demonstrate that the RNA in the final tube differed from the initial sample, and that it replicated faster than the initial sample. The RNA had evolved by the classical Darwinian mechanisms of mutation and natural selection.

Mistakes in copying had produced mutant RNA strands which competed for the supply of energy-rich precursor molecules (ATP, UTP, GTP and CTP). The most rapidly-reproducing mutants survived. Was Spiegelman's experiment merely a simulation of an early stage of biological evolution? Or was evolution of an extremely primitive life-form actually taking place in his test tubes?

G.F. Joyce, D.P. Bartel and others have performed experiments in which strands of RNA with specific catalytic activity (ribozymes) have been made to evolve artificially from randomly coded starting populations of RNA. In these experiments, starting populations of 10^{13} to 10^{15} randomly coded RNA molecules are tested for the desired catalytic activity, and the most successful molecules are then chosen as parents for the next generation. The selected molecules are replicated many times, but errors (mutations) sometimes occur in the replication. The new population is once again tested for catalytic activity, and the process is repeated. The fact that artificial evolution of ribozymes is possible can perhaps be interpreted as supporting the "RNA world" hypothesis, i.e., the hypothesis that RNA preceded DNA and proteins in the early history of terrestrial life.

In Chapter 4, we mentioned that John von Neumann speculated on the possibility of constructing artificial self-reproducing automata. In the early 1940's, a period when there was much discussion of the Universal Turing Machine, he became interested in constructing a mathematical model of the requirements for self-reproduction. Besides the Turing machine, another source of his inspiration was the paper by Warren McCulloch and Walter Pitts entitled *A logical calculus of the ideas immanent in nervous activity*, which von Neumann read in 1943. In his first attempt (the kinematic model), he imagined an extremely large and complex automaton, floating on a lake which contained its component parts.

Von Neumann's imaginary self-reproducing automaton consisted of four units, A, B, C and D. Unit A was a sort of factory, which gathered component parts from the surrounding lake and assembled them according to instructions which it received from other units. Unit B was a copying unit, which reproduced sets of instructions. Unit C was a control apparatus, similar to a computer. Finally D was a long string of instructions, analogous to the "tape" in the Turing machine described in Chapter 7. In von Neumann's kinematic automaton, the instructions were coded as a long binary number. The presence of what he called a "girder" at a given position corresponded to 1, while its absence corresponded to 0. In von Neumann's model, the automaton completed the assembly of its offspring by injecting its progeny with the duplicated instruction tape, thus making the new automaton both functional and fertile.

In presenting his kinematic model at the Hixton Symposium (organized by Linus Pauling in the late 1940's), von Neumann remarked that "...it is clear that the instruction [tape] is roughly effecting the function of a gene. It is also clear that the copying mechanism B performs the fundamental act of reproduction, the duplication of the genetic material, which is clearly the fundamental operation in the multiplication of living cells. It is also easy to see how arbitrary alterations of the system...can exhibit certain traits which appear in connection with mutation, lethality as a rule, but with a possibility of continuing reproduction with a modification of traits".

It is very much to von Neumann's credit that his kinematic model (which he invented several years before Crick and Watson published their DNA structure) was organized in much the same way that we now know the reproductive apparatus of a cell to be organized. Nevertheless he was dissatisfied with the model because his automaton contained too many "black boxes". There were too many parts which were supposed to have certain functions, but for which it seemed very difficult to propose detailed mechanisms by which the functions could be carried out. His kinematic model seemed very far from anything which could actually be built[8].

Von Neumann discussed these problems with his close friend, the Polish-American mathematician Stanislaw Ulam, who had for a long time been interested in the concept of self-replicating automata. When presented with the black box difficulty, Ulam suggested that the whole picture of an automaton floating on a lake containing its parts should be discarded. He proposed instead a model which later came to be known as the Cellular Automaton Model. In Ulam's model, the self-reproducing automaton lives in a very special space. For example, the space might resemble an infinite checkerboard, each square would constitute a multi-state cell. The state of each cell in a particular time interval is governed by the states of its near neighbors in the preceding time interval according to relatively simple laws. The automaton would then consist of a special configuration of cell states, and its reproduction would correspond to production of a similar

---

[8] Von Neumann's kinematic automaton was taken seriously by the Mission IV Group, part of a ten-week program sponsored by NASA in 1980 to study the possible use of advanced automation and robotic devices in space exploration. The group, headed by Richard Laing, proposed plans for self-reproducing factories, designed to function on the surface of the moon or the surfaces of other planets. Like von Neumann's kinetic automaton, to which they owed much, these plans seemed very far from anything that could actually be constructed.

configuration of cell states in a neighboring region of the cell lattice.

Von Neumann liked Ulam's idea, and he began to work in that direction. However, he wished his self-replicating automaton to be able to function as a universal Turing machine, and therefore the plans which he produced were excessively complicated. In fact, von Neumann believed complexity to be a necessary requirement for self-reproduction. In his model, the cells in the lattice were able to have 29 different states, and the automaton consisted of a configuration involving hundreds of thousands of cells. Von Neumann's manuscript on the subject became longer and longer, and he did not complete it before his early death from prostate cancer in 1957. The name "cellular automaton" was coined by Arthur Burks, who edited von Neumann's posthumous papers on the theory of automata.

Arthur Burks had written a Ph.D. thesis in philosophy on the work of the nineteenth century thinker Charles Sanders Pierce, who is today considered to be one of the founders of semiotics[9]. He then studied electrical engineering at the Moore School in Philadelphia, where he participated in the construction of ENIAC, one of the first general purpose electronic digital computers, and where he also met John von Neumann. He worked with von Neumann on the construction of a new computer, and later Burks became the leader of the Logic of Computers Group at the University of Michigan. One of Burks' students at Michigan was John Holland, the pioneer of genetic algorithms. Another student of Burks, E.F. Codd, was able to design a self-replicating automaton of the von Neumann type using a cellular automaton system with only 8 states (as compared with von Neumann's 29). For many years, enthusiastic graduate students at the Michigan group continued to do important research on the relationships between information, logic, complexity and biology.

Meanwhile, in 1968, the mathematician John Horton Conway, working in England at Cambridge University, invented a simple game which greatly increased the popularity of the cellular automaton concept. Conway's game, which he called "Life", was played on an infinite checker-board-like lattice of cells, each cell having only two states, "alive" or "dead". The rules which Conway proposed are as follows: "If a cell on the checkerboard is alive, it will survive in the next time step (generation) if there are either two or three neighbors also alive. It will die of overcrowding if there are more than three live neighbors, and it will die of exposure if there are fewer than two. If a cell on the checkerboard is dead, it will remain dead in the next

---

[9] Semiotics is defined as the study of signs (see Appendix 2).

generation unless exactly three of its eight neighbors is alive. In that case, the cell will be 'born' in the next generation".

Originally Conway's Life game was played by himself and by his colleagues at Cambridge University's mathematics department in their common room: At first the game was played on table tops at tea time. Later it spilled over from the tables to the floor, and tea time began to extend: far into the afternoons. Finally, wishing to convert a wider audience to his game, Conway submitted it to Martin Gardner, who wrote a popular column on "Mathematical Games" for the Scientific American. In this way Life spread to MIT's Artificial Intelligence Laboratory, where it created such interest that the MIT group designed a small computer specifically dedicated to rapidly implementing Life's rules.

The reason for the excitement about Conway's Life game was that it seemed capable of generating extremely complex patterns, starting from relatively simple configurations and using only its simple rules. Ed Fredkin, the director of MIT's Artificial Intelligence Laboratory, became enthusiastic about cellular automata because they seemed to offer a model for the way in which complex phenomena can emerge from the laws of nature, which are after all very simple. In 1982, Fredkin (who was independently wealthy because of a successful computer company which he had founded) organized a conference on cellular automata on his private island in the Caribbean. The conference is notable because one of the participants was a young mathematical genius named Stephen Wolfram, who was destined to refine the concept of cellular automata and to become one of the leading theoreticians in the field[10].

One of Wolfram's important contributions was to explore exhaustively the possibilities of 1-dimensional cellular automata. No one before him had looked at 1-dimensional CA's, but in fact they had two great advantages: The first of these advantages was simplicity, which allowed Wolfram to explore and classify the possible rule sets. Wolfram classified the rule sets into 4 categories, according to the degree of complexity which they generated. The second advantage was that the configurations of the system in successive generations could be placed under one another to form an easily-surveyed 2-dimensional visual display. Some of the patterns generated in this way were strongly similar to the patterns of pigmentation on the shells of certain molluscs. The strong resemblance seemed to suggest that Wolfram's 1-dimensional cellular automata might yield insights into

---

[10] As many readers probably know, Stephen Wolfram was also destined to become a millionaire by inventing the elegant symbol-manipulating program system, Mathematica.

the mechanism by which the pigment patterns are generated.

In general, cellular automata seemed to be promising models for gaining insight into the fascinating and highly important biological problem of morphogenesis: How does the fertilized egg translate the information on the genome into the morphology of the growing embryo, ending finally with the enormously complex morphology of a fully developed and fully differentiated multicellular animal? Our understanding of this amazing process is as yet very limited, but there is evidence that as the embryo of a multicellular animal develops, cells change their state in response to the states of neighboring cells. In the growing embryo, the "state" of a cell means the way in which it is differentiated, i.e., which genes are turned on and which off - which information on the genome is available for reading, and which segments are blocked. Neighboring cells signal to each other by means of chemical messengers[11]. Clearly there is a close analogy between the way complex patterns develop in a cellular automaton, as neighboring cells influence each other and change their states according to relatively simple rules, and the way in which the complex morphology of a multicellular animal develops in the growing embryo.

Conway's Life game attracted another very important worker to the field of cellular automata: In 1971, Christopher Langton was working as a computer programmer in the Stanley Cobb Laboratory for Psychiatric Research at Massachusetts General Hospital. When colleagues from MIT brought to the laboratory a program for executing Life, Langton was immediately interested. He recalls "It was the first hint that there was a distinction between the hardware and the behavior which it would support... You had the feeling that there was something very deep here in this little artificial universe and its evolution through time. [At the lab] we had a lot of discussions about whether the program could be open ended - could you have a universe in which life could evolve?"

Later, at the University of Arizona, Langton read a book describing von Neumann's theoretical work on automata. He contacted Arthur Burks, von Neumann's editor, who told him that no self-replicating automaton had actually been implemented, although E.F. Codd had proposed a simplified plan with only 8 states instead of 29. Burks suggested to Langton that he should start by reading Codd's book.

When Langton studied Codd's work, he realized that part of the problem was that both von Neumann and Codd had demanded that the self-

---

[11] We can recall the case of slime mold cells which signal to each other by means of the chemical messenger, cyclic AMP (Chapter 3).

reproducing automaton should be able to function as a universal Turing machine, i.e., as a universal computer. When Langton dropped this demand (which he considered to be more related to mathematics than to biology) he was able to construct a relatively simple self-reproducing configuration in an 8-state 2-dimensional lattice of CA cells. As they reproduced themselves, Langton's loop-like cellular automata filled the lattice of cells in a manner reminiscent of a growing coral reef, with actively reproducing loops on the surface of the filled area, and "dead" (nonreproducing) loops in the center.

Langton continued to work with cellular automata as a graduate student at Arthur Burks' Logic of Computers Group at Michigan. His second important contribution to the field was an extension of Wolfram's classification of rule sets for cellular automata. Langton introduced a parameter λ to characterize various sets of rules according to the type of behavior which they generated. Rule sets with a value near to the optimum ($\lambda = 0.273$) generated complexity similar to that found in biological systems. This value of Langton's λ parameter corresponded to a borderline region between periodicity and chaos.

After obtaining a Ph.D. from Burks' Michigan group, Christopher Langton moved to the Center for Nonlinear Studies at Los Alamos, New Mexico, where in 1987 he organized an "Interdisciplinary Workshop on the Synthesis and Simulation of Living Systems" - the first conference on artificial life ever held. Among the participants were Richard Dawkins, Astrid Lindenmayer, John Holland, and Richard Laing. The noted Oxford biologist and author Richard Dawkins was interested in the field because he had written a computer program for simulating and teaching evolution. Astrid Lindenmayer and her coworkers in Holland had written programs capable of simulating the morphogenesis of plants in an astonishingly realistic way. As was mentioned above, John Holland pioneered the development of genetic algorithms, while Richard Laing was the leader of NASA's study to determine whether self-reproducing factories might be feasible.

Langton's announcement for the conference, which appeared in the Scientific American, stated that "Artificial life is the study of artificial systems that exhibit behavior characteristic of natural living systems... The ultimate goal is to extract the logical form of living systems. Microelectronic technology and genetic engineering will soon give us the capability to create new life *in silico* as well as *in vitro*. This capacity will present humanity with the most far-reaching technical, theoretical, and ethical challenges it has ever confronted. The time seems appropriate for a gathering of those involved in attempts to simulate or synthesize aspects of living systems".

In the 1987 workshop on artificial life, a set of ideas which had gradually emerged during the previous decades of work on automata and simulations of living systems became formalized and crystallized: All of the participants agreed that something more than reductionism was needed to understand the phenomenon of life. This belief was not a revival of vitalism; it was instead a conviction that the abstractions of molecular biology are not in themselves sufficient. The type of abstraction found in Darwin's theory of natural selection was felt to be nearer to what was needed. The viewpoints of thermodynamics and statistical mechanics were also helpful. What was needed, it was felt, were insights into the flow of information in complex systems; and computer simulations could give us this insight. The fact that the simulations might take place in silico did not detract from their validity. The logic and laws governing complex systems and living systems were felt to be independent of the medium.

As Langton put it, "The ultimate goal of artificial life would be to create 'life' in some other medium, ideally a virtual medium where the essence of life has been abstracted from the details of its implementation in any particular model. We would like to build models that are so lifelike that they cease to become models of life and become examples of life themselves".

Most of the participants at the first conference on artificial life had until then been working independently, not aware that many other researchers shared their viewpoint. Their conviction that the logic of a system is largely independent of the medium echoes the viewpoint of the Macy Conferences on cybernetics in the 1940's, where the logic of feedback loops and control systems was studied in a wide variety of contexts, ranging from biology and anthropology to computer systems. A similar viewpoint can also be found in biosemiotics (Appendix 2), where, in the words of the Danish biologist Jesper Hoffmeyer, "the sign, rather than the molecule" is considered to be the starting point for studying life. In other words, the essential ingredient of life is information; and information can be expressed in many ways. The medium is less important than the message.

The conferences on artificial life have been repeated each year since 1987, and European conferences devoted to the new and rapidly growing field have also been organized. Langton himself moved to the Santa Fe Institute, where he became director of the institute's artificial life program and editor of a new journal, *Artificial Life*. The first three issues of the journal have been published as a book by the MIT Press, and the book presents an excellent introduction to the field.

Among the scientists who were attracted to the artificial life conferences was the biologist Thomas Ray, a graduate of Florida State University and Harvard, and an expert in the ecology of tropical rain forests. In the late 1970's, while he was working on his Harvard Ph.D., Ray happened to have a conversation with a computer expert from the MIT Artificial Intelligence Lab, who mentioned to him that computer programs can replicate. To Ray's question "How?", the AI man answered "Oh, it's trivial".

Ray continued to study tropical ecologies, but the chance conversation from his Cambridge days stuck in his mind. By 1989 he had acquired an academic post at the University of Delaware, and by that time he had also become proficient in computer programming. He had followed with interest the history of computer viruses. Were these malicious creations in some sense alive? Could it be possible to make self-replicating computer programs which underwent evolution by natural selection? Ray considered John Holland's genetic algorithms to be analogous to the type of selection imposed by plant and animal breeders in agriculture. He wanted to see what would happen to populations of digital organisms that found their own criteria for natural selection — not humanly imposed goals, but self-generated and open-ended criteria growing naturally out of the requirements for survival.

Although he had a grant to study tropical ecologies, Ray neglected the project and used most of his time at the computer, hoping to generate populations of computer organisms that would evolve in an open-ended and uncontrolled way. Luckily, before starting his work in earnest, Thomas Ray consulted Christopher Langton and his colleague James Farmer at the Center for Nonlinear Studies in New Mexico. Langton and Farmer realized that Ray's project could be a very dangerous one, capable of producing computer viruses or worms far more malignant and difficult to eradicate than any the world had yet seen. They advised Ray to make use of Turing's concept of a virtual computer. Digital organisms created in such a virtual computer would be unable to live outside it. Ray adopted this plan, and began to program a virtual world in which his freely evolving digital organisms could live. He later named the system "Tierra".

Ray's Tierra was not the first computer system to aim at open-ended evolution. Steen Rasmussen, working at the Danish Technical University, had previously produced a system called "VENUS" (Virtual Evolution in a Nonstochastic Universe Simulator) which simulated the very early stages of the evolution of life on earth. However, Ray's aim was not to understand the origin of life, but instead to produce digitally something analogous to the evolutionary explosion of diversity that occurred on earth at the start of the Cambrian era. He programmed an 80-byte self-reproducing digital organism which he called "Ancestor", and placed it in Tierra, his virtual Garden of Eden.

Ray had programmed a mechanism for mutation into his system, but he doubted that he would be able to achieve an evolving population with his first attempt. As it turned out, Ray never had to program another organism. His 80-byte Ancestor reproduced and populated his virtual earth, changing under the action of mutation and natural selection in a way that astonished and delighted him.

In his freely evolving virtual zoo, Ray found parasites, and even hyperparasites, but he also found instances of altruism and symbiosis. Most astonishingly of all, when he turned off the mutations in his Eden, his organisms invented sex (using mechanisms which Ray had introduced to allow for parasitism). They had never been told about sex by their creator, but they seemed to find their own way to the Tree of Knowledge.

Thomas Ray expresses the aims of his artificial life research as follows:[12] "Everything we know about life is based on one example: Life on Earth. Everything we know about intelligence is based on one example: Human intelligence. This limited experience burdens us with preconceptions, and limits our imaginations... How can we go beyond our conceptual limits, find the natural form of intelligent processes in the digital medium, and work with the medium to bring it to its full potential, rather than just imposing the world we know upon it by forcing it to run a simulation of our physics, chemistry and biology?..."

"In the carbon medium it was evolution that explored the possibilities inherent in the medium, and created the human mind. Evolution listens to the medium it is embedded in. It has the advantage of being mindless, and therefore devoid of preconceptions, and not limited by imagination."

"I propose the creation of a digital nature - a system of wildlife reserves in cyberspace in the interstices between human colonizations, feeding off unused CPU-cycles and permitted a share of our bandwidth. This would be a place where evolution can spontaneously generate complex information processes, free from the demands of human engineers and market analysts telling it what the target applications are - a place for a digital Cambrian explosion of diversity and complexity..."

"It is possible that out of this digital nature, there might emerge a

---

[12] T. Ray, http://www.hip.atr.co.jp/ ray/pubs/pubs.html

digital intelligence, truly rooted in the nature of the medium, rather than brutishly copied from organic nature. It would be a fundamentally alien intelligence, but one that would complement rather than duplicate our talents and abilities".

In Thomas Ray's experiments, the source of thermodynamic information is the electrical power needed to run the computer. In an important sense one might say that the digital organisms in Ray's Tierra system are living. This type of experimentation is in its infancy, but since it combines the great power of computers with the even greater power of natural selection, it is hard to see where it might end, and one can fear that it will end badly despite the precaution of conducting the experiments in a virtual computer.

Have Thomas Ray and other "a-lifers"[13] created artificial living organisms? Or have they only produced simulations that mimic certain aspects of life? Obviously the answer to this question depends on the definition of life, and there is no commonly agreed-upon definition. Does life have to involve carbon chemistry? The a-lifers call such an assertion "carbon chauvinism". They point out that elsewhere in the universe there may exist forms of life based on other media, and their program is to find medium-independent characteristics which all forms of life must have.

In the present book, especially in Chapter 4, we have looked at the phenomenon of life from the standpoint of thermodynamics, statistical mechanics and information theory. Seen from this viewpoint, a living organism is a complex system produced by an input of thermodynamic information in the form of Gibbs free energy. This incoming information keeps the system very far away from thermodynamic equilibrium, and allows it to achieve a statistically unlikely and complex configuration. The information content of any complex (living) system is a measure of how unlikely it would be to arise by chance. With the passage of time, the entropy of the universe increases, and the almost unimaginably improbable initial configuration of the universe is converted into complex free-energy-using systems that could never have arisen by pure chance. Life maintains itself and evolves by feeding on Gibbs free energy, that is to say, by feeding on the enormous improbability of the initial conditions of the universe.

All of the forms of artificial life that we have discussed derive their complexity from the consumption of free energy. For example, Spiegelman's evolving RNA molecules feed on the Gibbs free energy of the phosphate bonds of their precursors, ATP, GTP, UTP, and CTP. This free energy

---

[13] In this terminology, ordinary biologists are "b-lifers".

is the driving force behind artificial evolution which Spiegelman observed. In his experiment, thermodynamic information in the form of high-energy phosphate bonds is converted into cybernetic information.

Similarly, in the polymerase chain reaction, discussed in Chapter 3, the Gibbs free energy of the phosphate bonds in the precursor molecules ATP, TTP, GTP and CTP drives the reaction. With the aid of the enzyme DNA polymerase, the soup of precursors is converted into a highly improbable configuration consisting of identical copies of the original sequence. Despite the high improbability of the resulting configuration, the entropy of the universe has increased in the copying process. The improbability of the set of copies is less than the improbability of the high energy phosphate bonds of the precursors.

The polymerase chain reaction reflects on a small scale, what happens on a much larger scale in all living organisms. Their complexity is such that they never could have originated by chance, but although their improbability is extremely great, it is less than the still greater improbability of the configurations of matter and energy from which they arose. As complex systems are produced, the entropy of the universe continually increases, i.e., the universe moves from a less probable configuration to a more probable one.

Suggestions for further reading

(1) P. Priedland and L.H. Kedes, *Discovering the secrets of DNA*, Comm. of the ACM, **28**, 1164-1185 (1985).

(2) E.F. Meyer, *The first years of the protein data bank*, Protein Science **6**, 1591-7, July (1997).

(3) C. Kulikowski, *Artificial intelligence in medicine: History, evolution and prospects*, in Handbook of Biomedical Engineering, J. Bronzine editor, 181.1-181.18, CRC and IEEE Press, Boca Raton Fla., (2000).

(4) C. Gibas and P. Jambeck, *Developing Bioinformatics Computer Skills*, O'Reily, (2001).

(5) F.L. Carter, *The molecular device computer: point of departure for large-scale cellular automata*, Physica D, **10**, 175-194 (1984).

(6) K.E. Drexler, *Molecular engineering: an approach to the development of general capabilities for molecular manipulation*, Proc. Natl. Acad. Sci USA, **78**, 5275-5278 (1981).

(7) K.E. Drexler, *Engines of Creation*, Anchor Press, Garden City, New York, (1986).

(8) D.M. Eigler and E.K. Schweizer, *Positioning single atoms with a scanning electron microscope*, Nature, **344**, 524-526 (1990).

(9) E.D. Gilbert, editor, *Miniaturization*, Reinhold, New York, (1961).

(10) R.C. Haddon and A.A. Lamola, *The molecular electronic devices and the biochip computer: present status*, Proc. Natl. Acad. Sci. USA, **82**, 1874-1878 (1985).

(11) H.M. Hastings and S. Waner, *Low dissipation computing in biological systems*, BioSystems, **17**, 241-244 (1985).

(12) J.J. Hopfield, J.N. Onuchic and D.N. Beritan, *A molecular shift register based on electron transfer*, Science, **241**, 817-820 (1988).

(13) L. Keszthelyi, *Bacteriorhodopsin, in Bioenergetics*, P. P. Graber and G. Millazo (editors), Birkhäusr Verlag, Basil Switzerland, (1997).

(14) F.T. Hong, *The bacteriorhodopsin model membrane as a prototype molecular computing element*, BioSystems, **19**, 223-236 (1986).

(15) L.E. Kay, *Life as technology: Representing, intervening and molecularizing*, Rivista di Storia della Scienzia, **II**, 1, 85-103 (1993).

(16) A.P. Alivisatos et al., *Organization of 'nanocrystal molecules' using DNA*, Nature, **382**, 609-611, (1996).

(17) T. Bjørnholm et al., *Self-assembly of regioregular, amphiphilic polythiophenes into highly ordered pi-stacked conjugated thin films and nanocircuits*, J. Am. Chem. Soc. **120**, 7643 (1998).

(18) L.J. Fogel, A.J.Owens, and M.J. Walsh, *Artificial Intelligence Through Simulated Evolution*, John Wiley, New York, (1966).

(19) L.J. Fogel, *A retrospective view and outlook on evolutionary algorithms, in Computational Intelligence: Theory and Applications*, in *5th Fuzzy Days*, B. Reusch, editor, Springer-Verlag, Berlin, (1997).

(20) P.J. Angeline, *Multiple interacting programs: A representation for evolving complex behaviors*, Cybernetics and Systems, **29 (8)**, 779-806 (1998).

(21) X. Yao and D.B. Fogel, editors, *Proceedings of the 2000 IEEE Symposium on Combinations of Evolutionary Programming and Neural Networks*, IEEE Press, Piscataway, NJ, (2001).

(22) R.M. Brady, *Optimization strategies gleaned from biological evolution*, Nature **317**, 804-806 (1985).

(23) K. Dejong, *Adaptive system design — a genetic approach*, IEEE Syst. M. 10, 566-574 (1980).

(24) W.B. Dress, *Darwinian optimization of synthetic neural systems*, IEEE Proc. **ICNN 4**, 769-776 (1987).

(25) J.H. Holland, *A mathematical framework for studying learning in*

classifier systems, Physica **22 D**, 307-313 (1986).

(26) R.F. Albrecht, C.R. Reeves, and N.C. Steele (editors), *Artificial Neural Nets and Genetic Algorithms*, Springer Verlag, (1993).

(27) L. Davis, editor, *Handbook of Genetic Algorithms*, Van Nostrand Reinhold, New York, (1991).

(28) Z. Michalewicz, *Genetic Algorithms + Data Structures = Evolution Programs*, Springer-Verlag, New York, (1992), second edition, (1994).

(29) K.I. Diamantaris and S.Y. Kung, *Principal Component Neural Networks: Theory and Applications*, John Wiley and Sons, New York, (1996).

(30) A. Garliauskas and A. Soliunas, *Learning and recognition of visual patterns by human subjects and artificial intelligence systems*, Informatica, **9 (4)**, (1998).

(31) A. Garliauskas, *Numerical simulation of dynamic synapse-dendrite-soma neuronal processes*, Informatica, **9 (2)**, 141-160, (1998).

(32) U. Seifert and B. Michaelis, *Growing multi-dimensional self-organizing maps*, International Journal of Knowledge-Based Intelligent Engineering Systems,**2 (1)**, 42-48, (1998).

(33) S. Mitra, S.K. Pal, and M.K. Kundu, *Finger print classification using fuzzy multi-layer perceptron*, Neural Computing and Applications, **2**, 227-233 (1994).

(34) M. Verleysen (editor), *European Symposium on Artificial Neural Networks*, D-Facto, (1999).

(35) R.M. Golden, *Mathematical Methods for Neural Network Analysis and Design*, MIT Press, Cambridge MA, (1996).

(36) S. Haykin, *Neural Networks — (A) Comprehensive Foundation*, MacMillan, New York, (1994).

(37) M.A. Gronroos, *Evolutionary Design of Neural Networks*, Thesis, Computer Science, Department of Mathematical Sciences, University of Turku, Finland, (1998).

(38) D.E. Goldberg, *Genetic Algorithms in Search, Optimization and Machine Learning*, Addison-Wesley, (1989).

(39) M. Mitchell, An Introduction to Genetic Algorithms, MIT Press, Cambridge MA, (1996).

(40) L. Davis (editor), *Handbook of Genetic Algorithms*, Van Nostrand and Reinhold, New York, (1991).

(41) J.H. Holland, *Adaptation in Natural and Artificial Systems*, MIT Press, Cambridge MA, (1992).

(42) J.H. Holland, *Hidden Order; How Adaptation Builds Complexity*, Ad-

dison Wesley, (1995).

(43) W. Banzhaf, P. Nordin, R.E. Keller and F. Francone, *Genetic Programming - An Introduction; On the Automatic Evolution of Computer Programs and its Applications*, Morgan Kaufmann, San Francisco CA, (1998).

(44) W. Banzhaf et al. (editors), *(GECCO)-99: Proceedings of the Genetic Evolutionary Computation Conference*, Morgan Kaufman, San Francisco CA, (2000).

(45) W. Banzhaf, *Editorial Introduction, Genetic Programming and Evolvable Machines*, 1, 5-6, (2000).

(46) W. Banzhaf, *The artificial evolution of computer code, IEEE Intelligent Systems*, 15, 74-76, (2000).

(47) J.J. Grefenstette (editor), *Proceedings of the Second International Conference on Genetic Algorithms and their Applications*, Lawrence Erlbaum Associates, Hillsdale New Jersey, (1987).

(48) J. Koza, *Genetic Programming: On the Programming of Computers by means of Natural Selection*, MIT Press, Cambridge MA, (1992).

(49) J. Koza et al., editors, *Genetic Programming 1997: Proceedings of the Second Annual Conference*, Morgan Kaufmann, San Francisco, (1997).

(50) W.B. Langdon, *Genetic Programming and Data Structures*, Kluwer, (1998).

(51) D. Lundh, B. Olsson, and A. Narayanan, editors, *Bio-Computing and Emergent Computation 1997*, World Scientific, Singapore, (1997).

(52) P. Angeline and K. Kinnear, editors, *Advances in Genetic Programming: Volume 2*, MIT Press, (1997).

(53) J.H. Holland, *Adaptation in Natural and Artificial Systems, The University of Michigan Press*, Ann Arbor, (1975).

(54) David B. Fogel and Wirt Atmar (editors), *Proceedings of the First Annual Conference on Evolutionary Programming, Evolutionary Programming Society*, La Jolla California, (1992).

(55) M. Sipper et al., *A phylogenetic, ontogenetic, and epigenetic view of bioinspired hardware systems*, IEEE Transactions in Evolutionary Computation 1, 1 (1997).

(56) E. Sanchez and M. Tomassini, editors, *Towards Evolvable Hardware*, Lecture Notes in Computer Science, 1062, Springer-Verlag, (1996).

(57) J. Markoff, *A Darwinian creation of software*, New York Times, Section C, p.6, February 28, (1990).

(58) A. Thompson, *Hardware Evolution: Automatic design of electronic*

circuits in reconfigurable hardware by artificial evolution, Distinguished dissertation series, Springer-Verlag, (1998).

(59) W. McCulloch and W. Pitts, *A Logical Calculus of the Ideas Immanent in Nervous Activity*, Bulletin of Mathematical Biophysics, **7**, 115-133, (1943).

(60) F. Rosenblatt, *Principles of Neurodynamics*, Spartan Books, (1962).

(61) C. von der Malsburg, *Self-Organization of Orientation Sensitive Cells in the Striate Cortex*, Kybernetik, **14**, 85-100, (1973).

(62) S. Grossberg, *Adaptive Pattern Classification and Universal Recoding: 1. Parallel Development and Coding of Neural Feature Detectors*, Biological Cybernetics, **23**, 121-134, (1976).

(63) J.J. Hopfield and D.W. Tank, *Computing with Neural Circuits: A Model*, Science, **233**, 625-633, (1986).

(64) R.D. Beer, *Intelligence as Adaptive Behavior: An Experiment in Computational Neuroethology*, Academic Press, New York, (1990).

(65) S. Haykin, *Neural Networks: A Comprehensive Foundation*, IEEE Press and Macmillan, (1994).

(66) S.V. Kartalopoulos, *Understanding Neural Networks and Fuzzy Logic: Concepts and Applications*, IEEE Press, (1996).

(67) D. Fogel, *Evolutionary Computation: The Fossil Record*, IEEE Press, (1998).

(68) D. Fogel, *Evolutionary Computation: Toward a New Philosophy of Machine Intelligence*, IEEE Press, Piscataway NJ, (1995).

(69) J.M. Zurada, R.J. Marks II, and C.J. Robinson, editors, *Computational Intelligence: Imitating Life*, IEEE Press, (1994).

(70) J. Bezdek and S.K. Pal, editors, *Fuzzy Models for Pattern Recognition: Methods that Search for Structure in Data*, IEEE Press, (1992).

(71) M.M. Gupta and G.K. Knopf, editors, *Neuro-Vision Systems: Principles and Applications*, IEEE Press, (1994).

(72) C. Lau, editor, *Neural Networks. Theoretical Foundations and Analysis*, IEEE Press, (1992).

(73) T. Back, D.B. Fogel and Z. Michalewicz, editors, *Handbook of Evolutionary Computation*, Oxford University Press, (1997).

(74) D.E. Rumelhart and J.L. McClelland, *Parallel Distributed Processing: Explorations in the Micro structure of Cognition, Volumes I and II*, MIT Press, (1986).

(75) J. Hertz, A. Krogh and R.G. Palmer, *Introduction to the Theory of Neural Computation*, Addison Wesley, (1991).

(76) J.A. Anderson and E. Rosenfeld, *Neurocomputing: Foundations of*

Research, MIT Press, (1988).

(77) R.C. Eberhart and R.W. Dobbins, *Early neural network development history: The age of Camelot*, IEEE Engineering in Medicine and Biology **9**, 15-18 (1990).

(78) T. Kohonen, *Self-Organization and Associative Memory*, Springer-Verlag, Berlin, (1984).

(79) T. Kohonen, *Self-Organizing Maps*, Springer-Verlag, Berlin, (1997).

(80) G.E. Hinton, *How neural networks learn from experience*, Scientific American **267**, 144-151 (1992).

(81) K. Swingler, *Applying Neural Networks: A Practical Guide*, Academic Press, New York, (1996).

(82) B.K. Wong, T.A. Bodnovich and Y. Selvi, *Bibliography of neural network business applications research: 1988-September 1994*, Expert Systems **12**, 253-262 (1995).

(83) I. Kaastra and M. Boyd, *Designing neural networks for forecasting financial and economic time series*, Neurocomputing **10**, 251-273 (1996).

(84) T. Poddig and H. Rehkugler, *A world model of integrated financial markets using artificial neural networks*, Neurocomputing **10**, 2251-273 (1996).

(85) J.A. Burns and G.M. Whiteside, *Feed forward neural networks in chemistry: Mathematical systems for classification and pattern recognition*, Chem. Rev. **93**, 2583-2601, (1993).

(86) M.L. Action and P.W. Wilding, *The application of backpropagation neural networks to problems in pathology and laboratory medicine*, Arch. Pathol. Lab. Med. **116**, 995-1001 (1992).

(87) D.J. Maddalena, *Applications of artificial neural networks to problems in quantitative structure activity relationships*, Exp. Opin. Ther. Patents **6**, 239-251 (1996).

(88) W.G. Baxt, *Application of artificial neural networks to clinical medicine*, [Review], Lancet **346**, 1135-8 (1995).

(89) A. Chablo, *Potential applications of artificial intelligence in telecommunications*, Technovation **14**, 431-435 (1994).

(90) D. Horwitz and M. El-Sibaie, *Applying neural nets to railway engineering*, AI Expert, 36-41, January (1995).

(91) J. Plummer, *Tighter process control with neural networks*, 49-55, October (1993).

(92) T. Higuchi et al., *Proceedings of the First International Conference on Evolvable Systems: From Biology to Hardware (ICES96)*, Lecture

Notes on Computer Science, Springer-Verlag, (1997).

(93) S.A. Kaufman, *Antichaos and adaption*, Scientific American, **265**, 78-84, (1991).

(94) S.A. Kauffman, *The Origins of Order*, Oxford University Press, (1993).

(95) M.M. Waldrop, *Complexity: The Emerging Science at the Edge of Order and Chaos*, Simon and Schuster, New York, (1992).

(96) H.A. Simon, *The Science of the Artificial, 3rd Edition*, MIT Press, (1996).

(97) M.L. Hooper, *Embryonic Stem Cells: Introducing Planned Changes into the Animal Germline*, Harwood Academic Publishers, Philadelphia, (1992).

(98) F. Grosveld, (editor), *Transgenic Animals*, Academic Press, New York, (1992).

(99) G. Kohler and C. Milstein, *Continuous cultures of fused cells secreting antibody of predefined specificity*, Nature, **256**, 495-497 (1975).

(100) S. Spiegelman, *An approach to the experimental analysis of precellular evolution*, Quarterly Reviews of Biophysics, **4**, 213-253 (1971).

(101) M. Eigen, *Self-organization of matter and the evolution of biological macromolecules*, Naturwissenschaften, **58**, 465-523 (1971).

(102) M. Eigen and W. Gardiner, *Evolutionary molecular engineering based on RNA replication*, Pure and Applied Chemistry, **56**, 967-978 (1984).

(103) G.F. Joyce, *Directed molecular evolution*, Scientific American **267** (**6**), 48-55 (1992).

(104) N. Lehman and G.F. Joyce, *Evolution in vitro of an RNA enzyme with altered metal dependence*, Nature, **361**, 182-185 (1993).

(105) E. Culotta, *Forcing the evolution of an RNA enzyme in the test tube*, Science, **257**, 31 July, (1992).

(106) S.A. Kauffman, *Applied molecular evolution*, Journal of Theoretical Biology, **157**, 1-7 (1992).

(107) H. Fenniri, *Combinatorial Chemistry. A Practical Approach*, Oxford University Press, (2000).

(108) P. Seneci, *Solid-Phase Synthesis and Combinatorial Technologies*, John Wiley & Sons, New York, (2001).

(109) G.B. Fields, J.P. Tam, and G. Barany, *Peptides for the New Millennium*, Kluwer Academic Publishers, (2000).

(110) Y.C. Martin, *Diverse viewpoints on computational aspects of molecular diversity*, Journal of Combinatorial Chemistry, **3**, 231-250, (2001).

(111) C.G. Langton et al., editors, *Artificial Life II: Proceedings of the Workshop on Artificial Life Held in Santa Fe, New Mexico*, Addison-Wesley, Reading MA, (1992).

(112) W. Aspray and A. Burks, eds., *Papers of John von Neumann on Computers and Computer Theory*, MIT Press, (1967).

(113) M. Conrad and H.H. Pattee, *Evolution experiments with an artificial ecosystem*, J. Theoret. Biol., **28**, (1970).

(114) C. Emmeche, *Life as an Abstract Phenomenon: Is Artificial Life Possible?*, in *Toward a Practice of Artificial Systems: Proceedings of the First European Conference on Artificial Life*, MIT Press, Cambridge MA, (1992).

(115) C. Emmeche, *The Garden in the Machine: The Emerging Science of Artificial Life*, Princeton University Press, Princeton NJ, (1994).

(116) S. Levy, *Artificial Life: The Quest for New Creation*, Pantheon, New York, (1992).

(117) K. Lindgren and M.G. Nordahl, *Cooperation and Community Structure in Artificial Ecosystems*, Artificial Life, **1**, 15-38 (1994).

(118) P. Husbands and I. Harvey (editors), *Proceedings of the 4th Conference on Artificial Life (ECAL '97)*, MIT Press, (1997).

(119) C.G. Langton, (editor), *Artificial Life: An Overview*, MIT Press, Cambridge MA, (1997).

(120) C.G. Langton, ed., *Artificial Life*, Addison-Wesley, (1987).

(121) A.A. Beaudry and G.F. Joyce, *Directed evolution of an RNA enzyme*, Science, **257**, 635-641 (1992).

(122) D.P. Bartel and J.W. Szostak, *Isolation of new ribozymes from a large pool of random sequences*, Science, 261, 1411-1418 (1993).

(123) K. Kelly, *Out of Control*, www.kk.org/outofcontrol/index.html, (2002).

(124) K. Kelly, *The Third Culture*, Science, February 13, (1998).

(125) S. Blakeslee, *Computer life-form "mutates" in an evolution experiment, natural selection is found at work in a digital world*, New York Times, November 25, (1997).

(126) M. Ward, *It's life, but not as we know it*, New Scientist, July 4, (1998).

(127) P. Guinnessy, *"Life" crawls out of the digital soup*, New Scientist, April 13, (1996).

(128) L. Hurst and R. Dawkins, *Life in a test tube*, Nature, May 21, (1992).

(129) J. Maynard Smith, *Byte-sized evolution*, Nature, February 27, (1992).

(130) W.D. Hillis, *Intelligence as an Emergent Behavior*, in *Artificial In-*

telligence, S. Graubard, ed., MIT Press, (1988).

(131) T.S. Ray, *Evolution and optimization of digital organisms*, in *Scientific Excellence in Supercomputing: The IBM 1990 Contest Prize Papers*, K.R. Billingsly, E. Derohanes, and H. Brown, III, editors, The Baldwin Press, University of Georgia, Athens GA 30602, (1991).

(132) S. Lloyd, *The calculus of intricacy*, The Sciences, October, (1990).

(133) M. Minsky, *The Society of Mind*, Simon and Schuster, (1985).

(134) D. Pines, ed., *Emerging Synthesis in Science*, Addison-Wesley, (1988).

(135) P. Prusinkiewicz and A. Lindenmayer, *The Algorithmic Beauty of Plants*, Springer-Verlag, (1990).

(136) T. Tommaso and N. Margolus, *Cellular Automata Machines: A New Environment for Modeling*, MIT Press, (1987).

(137) W.M. Mitchell, *Complexity: The Emerging Science at the Edge of Order and Chaos*, Simon and Schuster, (1992).

(138) T.S. Ray et al., *Kurtzweil's Turing Fallacy, in Are We Spiritual Machines?: Ray Kurzweil vs. the Critics of Strong AI*, J. Richards, ed., Viking, (2002).

(139) T.S. Ray, *Aesthetically Evolved Virtual Pets*, in *Artificial Life 7 Workshop Proceedings*, C.C. Maley and E. Bordreau, eds., (2000).

(140) T.S. Ray and J.F. Hart, *Evolution of Differentiation in Digital Organisms*, in *Artificial Life VII, Proceedings of the Seventh International Conference on Artificial Life*, M.A. Bedau, J.S. McCaskill, N.H. Packard, and S. Rasmussen, eds., MIT Press, (2000).

(141) T.S. Ray, *Artificial Life, in Frontiers of Life, Vol. 1: The Origins of Life*, R. Dulbecco et al., eds., Academic Press, (2001).

(142) T.S. Ray, *Selecting naturally for differentiation: Preliminary evolutionary results*, Complexity, **3 (5)**, John Wiley and Sons, (1998).

(143) K. Sims, *Artificial Evolution for Computer Graphics*, Computer Graphics, **25 (4)**, 319-328 (1991).

(144) K. Sims, Galapagos, http://web.genarts.com/galapagos, (1997).

Chapter 9

LOOKING TOWARDS THE FUTURE

Tensions created by the rapidity of technological change

In human cultural evolution, information transfer and storage through the language of molecular complementarity is supplemented by new forms of biological information flow and conservation — spoken language, writing, printing, and more recently electronic communication. The result has been a shift into a much higher evolutionary gear.

Because of new, self-reinforcing mechanisms of information flow and accumulation, the rate of evolutionary change has increased enormously: It took 3 billion years for the first autocatalytic systems to develop into multicellular organisms. Five hundred million years were required for multicellular organisms to rise from the level of sponges and slime molds to the degree of complexity and organization that characterizes primates and other mammals; but when a branch of the primate family developed a tool-using culture, spoken language, and an enlarged brain, only 40,000 years were required for our ancestors to change from animal-like hunter-gatherers into engineers, poets and astronomers.

During the initial stages of human cultural evolution, the rate of change was slow enough for genetic adaptation to keep pace. The co-evolution of speech, tool use, and an enlarged brain in hominids took place over a period of several million years, and there was ample time for genetic adaptation. The prolonged childhood which characterizes our species, and the behavior patterns of familial and tribal solidarity, were built into the genomes of our ancestors during the era of slow change, when cultural and genetic evolution moved together in equilibrium. However, as the pace of cultural information accumulation quickened, genetic change could no longer keep up.

Genetically we are almost identical with our neolithic ancestors; but their world has been replaced by a world of quantum theory, relativity, supercomputers, antibiotics, genetic engineering and space telescopes — unfortunately also a world of nuclear weapons and nerve gas. Because of the slowness of genetic evolution in comparison to the rapid and constantly-accelerating rate of cultural change, our bodies and minds are not perfectly adapted to our new way of life. They reflect more accurately the way of life of our hunter-gatherer ancestors.

In addition to the contrast between the slow pace of genetic evolution when compared with the rapid and constantly-accelerating rate of cultural evolution, we can also notice a contrast between rapidly- and slowly-moving aspects of cultural change: Social institutions and structures seem to change slowly when compared with the lightning-like pace of scientific and technological innovation. Thus, tensions and instability characterize information-driven society, not only because science and technology change so much more rapidly than institutions, laws, and attitudes, but also because human nature is not completely appropriate to our present way of life. In particular, human nature seems to contain an element of what might be called "tribalism", because our emotions evolved during an era when our ancestors lived in small, mutually hostile tribes, competing with one another for territory on the grasslands of Africa.

Looking towards the future, what can we predict? Detailed predictions are very difficult, but it seems likely that information technology and biotechnology will for some time continue to be the most rapidly-developing branches of science, and that these two fields will merge. We can guess with reasonable certainty that much progress will be made in understanding the mechanism of the brain, and in duplicating its functions artificially. Scientists of the future will undoubtedly achieve greatly increased control over the process of evolution. Thus it seems probable that the rapidity of scientific and technological change will produce ethical dilemmas and social tensions even more acute than those which we experience today. It is likely that the fate of our species (and the fate of the biosphere) will be made precarious by the astonishing speed of scientific and technological change unless this progress is matched by the achievement of far greater ethical and political maturity than we have yet attained.

Science has proved to be double-edged — capable of great good, but also of great harm. Information-driven human cultural evolution is a spectacular success — but can it become stable? Terrestrial life can look back on almost four billion years of unbroken evolutionary progress. Can we say with confidence that an equal period stretches ahead of us?

Can information-driven society achieve stability?

"We are living in a very special time", Murray Gell-Mann[1] remarked in a recent interview, "Historians hate to hear this, because they have heard it so many times before, but we *are* living in a very special time. One symptom of this is the fact that human population has for a long time been increasing according to a hyperbolic curve — a constant divided by 2020 minus the year".

The graph of global human population as a function of time, to which Gell-Mann refers in this quotation, is shown in Figure 6.1. Estimates of population are indicated by dots on the graph, while the smooth curve shows the hyperbola $P = C/(2020 - y)$, P being the population, y, the year, and C a constant. The form of the smooth curve, which matches the dots with reasonable accuracy, is at first surprising. One might have expected it to be an exponential, if the rate of increase were proportional to the population already present. The fact that the curve is instead a hyperbola can be understood in terms of the accumulation of cultural information. New techniques (for example the initial invention of agriculture, the importation of potatoes to Europe, or the introduction of high-yield wheat and rice varieties) make population growth possible. In the absence of new techniques, population is usually held in check by the painful Malthusian forces — famine, disease, and war.

The curve in Figure 6.1 shows an explosive growth of human population, driven by an equally explosive growth of stored cultural information — especially agricultural and medical information, and the information needed for opening new land to agriculture. As Gell-Mann remarks, population cannot continue to increase in this way, because we are rapidly approaching the limits of the earth's carrying capacity. Will human numbers overshoot these limits and afterwards crash disastrously? There is certainly a danger that this will happen.

Besides the challenge of stabilizing global population, the information-driven human society of the future will face another daunting task: Because of the enormously destructive weapons that have already been produced through the misuse of science, and because of the even worse weapons that may be invented in the future, the long-term survival of civilization can only be insured if society is able to eliminate the institution of war. This task will be made more difficult by the fact that human nature seems to

---

[1]Gell-Mann is an American physicist who was awarded a Nobel Prize in 1969 for his contributions to the theory of elementary particles.

contain an element of tribalism.

Humans tend to show great kindness towards close relatives and members of their own group, and are even willing to sacrifice their lives in battle in defense of their own family, tribe or nation. This tribal altruism is often accompanied by inter-tribal aggression — great cruelty towards the "enemy", i.e., towards members of a foreign group which is perceived to be threatening ones own. The fact that human nature seems to contain a genetically-programmed tendency towards tribalism is the reason why we find football matches entertaining, and the reason why Arthur Koestler once remarked: "We can control the movements of a space-craft orbiting about a distant planet, but we cannot control the situation in Northern Ireland."

How could evolutionary forces have acted to make the pattern of tribal altruism and inter-tribal aggression a part of human nature? To put the same question differently, how could our ancestors have increased the chances for survival of their own genes by dying in battle? The statistician R.A. Fisher and the evolutionary biologist J.B.S. Haldane considered this question in the 1920's[2]. Their solution was the concept of population genetics, in which the genetically homogeneous group as a whole — now sometimes called the "deme" — is taken to be the unit upon which evolutionary forces act.

Haldane and Fisher postulated that the small tribes in which our ancestors lived were genetically homogeneous, since marriage within the tribe was more probable than marriage outside it. This being the case, a patriotic individual who died for the tribe, killing many members of a competing tribe in the process, increased the chance of survival for his or her own genes, which were carried into the future by the surviving members of the hero's group. The tribe as a whole either lived or died; and those with the best "team spirit" survived most frequently.

Because of the extraordinarily bitter and cruel conflicts between ethnic groups which can be found in both ancient and modern history, it is necessary to take the ideas of Haldane and Fischer seriously. This does not mean that the elimination of the institution of war is impossible, but it means that the task will require the full resources and full cooperation of the world's educational systems, religions, and mass media. It will be necessary to educate children throughout the world in such a way that they will think of humanity as a single group — a large family to which all humans

---

[2] More recently the evolution of tribal altruism and inter-tribal aggression has also been discussed by W.D. Hamilton and Richard Dawkins.

belong, and to which they owe their ultimate loyalty.

In addition to educational reform, and reform of the images presented by the mass media, the elimination of war will require the construction of a democratic, just, and humane system of international governance, whose laws will act on individuals rather than on states. The problems involved are very difficult, but they must be solved if the information-driven society of the future is to achieve stability.

Respect for natural evolution

The avalanche of new techniques in biotechnology and information technology will soon give scientists so much power over evolution that evolutionary ethical problems will become much more acute than they are today. It is already possible to produce chimeras, i.e., transgenic animals and plants incorporating genetic information from two or more species. Will we soon produce hybrids which are partly machines and partly living organisms? What about artificial life? Will humans make themselves obsolete by allowing far more intelligent beings to evolve in cyberspace, as Thomas Ray proposes? What about modification and improvement of our own species? Is there a limit beyond which we ought not to go in constructing new organisms to suit human purposes?

Perhaps one answer to these questions can be found by thinking of the way in which evolution has operated to produce the biosphere. Driven by the flood of Gibbs free energy which the earth receives from the sun, living organisms are generated and tested by life. New generations are randomly modified by the genetic lottery, sometimes for the worse, and sometimes for the better; and the instances of improvement are kept. It would be hard to overestimate the value of this mechanism of design by random modification and empirical testing, with the preservation of what works. The organisms which are living today are all champions! They are distillations of vast quantities of experience, end products of four billion years of solar energy income.

The beautiful and complex living organisms of our planet are exquisitely adapted to survive, to live with each other, and to form harmonious ecological systems. Whatever we do in biotechnology ought to be guided by caution and by profound respect for what evolution has already achieved. We need a sense of evolutionary responsibility, and a non-anthropocentric component in our system of ethics.

Construction versus destruction

It is often said that ethical principles cannot be derived from science — that they must come from somewhere else. Nevertheless, when nature is viewed through the eyes of modern science, we obtain some insights which seem almost ethical in character. Biology at the molecular level has shown us the complexity and beauty of even the most humble living organisms, and the interrelatedness of all life on earth. Looking through the eyes of contemporary biochemistry, we can see that even the single cell of an amoeba is a structure of miraculous complexity and precision, worthy of our respect and wonder.

Knowledge of the second law of thermodynamics — the statistical law favoring disorder over order — reminds us that life is always balanced like a tight-rope walker over an abyss of chaos and destruction. Living organisms distill their order and complexity from the flood of thermodynamic information which reaches the earth from the sun. In this way, they create local order; but life remains a fugitive from the second law of thermodynamics. Disorder, chaos, and destruction remain statistically favored over order, construction, and complexity.

It is easier to burn down a house than to build one, easier to kill a human than to raise and educate one, easier to force a species into extinction than to replace it once it is gone, easier to burn the Great Library of Alexandria than to accumulate the knowledge that once filled it, and easier to destroy a civilization in a thermonuclear war than to rebuild it from the radioactive ashes. Knowing this, scientists can form an almost ethical insight: To be on the side of order, construction, and complexity, is to be on the side of life. To be on the side of destruction, disorder, chaos and war is to be against life, a traitor to life, an ally of death. Knowing the precariousness of life — knowing the statistical laws that favor disorder and chaos, we should resolve to be loyal to the principle of long continued construction upon which life depends.

Suggestions for further reading

(1) D.F. Noble, *Forces of Production: A Social History of Industrial Automation*, Knopf, New York, (1984).
(2) E. Morgan, *The Scars of Evolution*, Oxford University Press, (1990).
(3) W.D. Hamilton, *The genetical theory of social behavior. I and II*, J. Theor. Biol. **7**, 1-52 (1964).

(4) R.W. Sussman, *The Biological Basis of Human Behavior*, Prentice Hall, Englewood Cliffs, (1997).

(5) H. von Foerster, *KybernEthik*, Merve Verlag, Berlin, (1993).

(6) L. Westra, *An Environmental Proposal for Ethics: The Principle of Integrity*, Rowman and Littlefield, Lanham MD, (1994).

(7) M. Murphy and L. O'Neill, editors, *What is Life? The Next Fifty Years: Speculations on the Future of Biology*, Cambridge University Press, (1997).

(8) Konrad Lorenz, *On Aggression*, Bantam Books, New York (1977).

(9) I. Eibl-Eibesfeldt, *The Biology of Peace and War*, Thames and Hudson, New York (1979).

(10) R.A. Hinde, *Biological Bases of Human Social Behavior*, McGraw-Hill, New York (1977).

(11) R.A. Hinde, *Towards Understanding Relationships*, Academic Press, London (1979).

(12) Albert Szent-Györgyi, *The Crazy Ape*, Philosophical Library, New York (1970).

(13) E.O. Wilson, *Sociobiology*, Harvard University Press (1975).

(14) C. Zhan-Waxier, *Altruism and Aggression: Biological and Social Origins*, Cambridge University Press (1986).

(15) R. Axelrod, *The Evolution of Cooperation*, Basic Books, New York (1984).

(16) B. Mazlish, *The Fourth Discontinuity: The Coevolution of Humans and Machines*, Yale University Press, (1993).

Appendix A

ENTROPY AND INFORMATION

In Chapter 4, we mentioned that Boltzmann was able to establish a relationship between entropy and missing information. In this appendix, we will look in detail at his reasoning.

The reader will remember that Boltzmann's statistical mechanics (seen from a modern point of view) deals with an ensemble of N weakly-interacting identical systems which may be in one or another of a set of discrete states, $i = 1, 2, 3, \ldots$ with energies ϵ_i, with the number of the systems which occupy a particular state denoted by n_i,

$$
\begin{array}{lllllll}
\text{State number} & 1 & 2 & 3 & \ldots i & \ldots \\
\\
\text{Energy} & \epsilon_1, & \epsilon_2, & \epsilon_3, & \ldots \epsilon_i, & \ldots \\
\\
\text{Occupation number} & n_1, & n_2, & n_3, & \ldots n_i, & \ldots
\end{array}
\tag{A.1}
$$

A "macrostate" of the N identical systems can be specified by writing down the energy levels and their occupation numbers. This macrostate can be constructed in many ways, and each of these ways is called a "microstate". From combinatorial analysis it is possible to show that the number of microstates corresponding to a given macrostate is given by:

$$
W = \frac{N!}{n_1! n_2! n_3! \ldots n_i! \ldots}
\tag{A.2}
$$

Boltzmann assumed that for very large values of N, the most probable macrostate predominates over all others. He also assumed that the amount of energy which is shared by the N identical systems has a constant value, E, so that

$$
\sum_i n_i \epsilon_i - E = 0
\tag{A.3}
$$

He knew, in addition, that the sum of the occupation numbers must be equal to the number of weakly-interacting identical systems:

$$\sum_i n_i - N = 0 \qquad (A.4)$$

It is logical to assume that all microstates which fulfill these two conditions are equally probable, since the N systems are identical. It then follows that the probability of a particular macrostate is proportional to the number of microstates from which it can be constructed, i.e., proportional to W, so that if we wish to find the most probable macrostate, we need to maximize W subject to the constraints (A.3) and (A.4). It turns out to be more convenient to maximize $\ln W$ subject to these two constraints, but maximizing In W will of course also maximize W. Using the method of undetermined Lagrange multipliers, we look for an absolute maximum of the function

$$\ln W - \lambda \left(\sum_i n_i - N \right) - \beta \left(\sum_i n_i \epsilon_i - E \right) \qquad (A.5)$$

Having found this maximum, we can use the conditions (A.3) and (A.4) to determine the values of the Lagrangian multipliers λ and β. For the function shown in equation (A.5) to be a maximum, it is necessary that its partial derivative with respect to each of the occupation numbers shall vanish. This gives us the set of equations

$$\frac{\partial}{\partial n_i} \left[\ln(N!) - \sum_i \ln(n_i) \right] - \lambda - \beta \epsilon_i = 0 \qquad (A.6)$$

which must hold for all values of i. For very large values of N and n_i, Sterling's approximation,

$$\ln(n_i) = n_i (\ln n_i - 1) \qquad (A.7)$$

can be used to simplify the calculation. With the help of Sterling's approximation and the identity

$$\frac{\partial}{\partial n_i} [n_i (\ln n_i - 1)] = \ln n_i \qquad (A.8)$$

we obtain the relationship

$$- \ln n_i - \lambda - \beta \epsilon_i = 0 \qquad (A.9)$$

which can be rewritten in the form

$$n_i = e^{-\lambda - \beta \epsilon_i} \qquad (A.10)$$

and for the most probable macrostate, this relationship must hold for all values of i. Substituting (A.10) into (A.4), we obtain:

$$N = \sum_i n_i = e^{-\lambda} \sum_i e^{-\beta\epsilon_i} \tag{A.11}$$

so that

$$\frac{n_i}{N} = \frac{e^{-\beta\epsilon_i}}{\sum_i e^{-\beta\epsilon_i}} \equiv \frac{e^{-\beta\epsilon_i}}{Z} \tag{A.12}$$

where

$$Z \equiv \sum_i e^{-\beta\epsilon_i} \tag{A.13}$$

The sum Z is called the "partition function" (or in German, Zustandssumme) of a system, and it plays a very central role in statistical mechanics. All of the thermodynamic functions of a system can be derived from it. The factor $e^{-\beta\epsilon_i}$ is called the "Boltzmann factor". Looking at equation (A.12), we can see that because of the Boltzmann factor, the probability

$$P_i \equiv \frac{n_i}{N} = \frac{e^{-\beta\epsilon_i}}{Z} \tag{A.14}$$

that a particular system will be in a state i is smaller for the states of high energy than it is for those of lower energy. We mentioned above that the constraints (A.3) and (A.4) can be used to find the values of the Lagrangian multipliers λ and β. The condition

$$E = N \sum_i P_i \epsilon_i \tag{A.15}$$

can be used to determine β. By applying his statistical methods to a monatomic gas at low pressure, Boltzmann found that

$$\beta = \frac{1}{kT} \tag{A.16}$$

where T is the absolute temperature and k is the constant which appears in the empirical law relating the pressure, volume and temperature of a perfect gas:

$$PV = nkT \tag{A.17}$$

From experiments on monatomic gases at low pressures, one finds that the "Boltzmann constant" k is given by

$$k = 1.38062 \times 10^{-23} \; \frac{\text{Joules}}{\text{Kelvin}} \tag{A.18}$$

We mentioned that Boltzmann's equation relating entropy to disorder is carved on his tombstone. With one minor difference, this equation is

$$S_N = k \ln W \tag{A.19}$$

(The minor difference is that on the tombstone, the S lacks a subscript.) How did Boltzmann identify $k \ln W$ with the entropy of Clausius, $dS = dq/T$? In answering this question we will continue to use modern picture of a system with a set of discrete states i, whose energies are ϵ_i. Making use of Sterling's approximation, equation (A.9), and remembering the definition of W, (A.2), we can rewrite (A.19) as

$$S_N = k \ln \left[\frac{N!}{n_1! n_2! n_3! ... n_i! ...} \right]$$

$$= k \left[\ln(N!) - \sum_i \ln(n_i)! \right] \approx -kN \sum_i \frac{n_i}{N} \ln \frac{n_i}{N} \tag{A.20}$$

Equation (A.20) gives us the entropy of the entire collection of N identical weakly-interacting systems. The entropy of a single system is just this quantity divided by N:

$$S = \frac{S_N}{N} = -k \sum_i P_i \ln P_i \equiv -k \langle lnP \rangle \tag{A.21}$$

where $P_i = n_i/N$, defined by equation (A.14), is the probability that the system is in state i. According to equation (A.14), this probability is just equal to the Boltzmann factor, $e^{-\beta \epsilon_i}$, divided by the partition function, Z, so that

$$S = -k \sum_i \frac{e^{-\beta \epsilon_i}}{Z} \ln \left(\frac{e^{-\beta \epsilon_i}}{Z} \right)$$

$$= \frac{k}{Z} \sum_i e^{-\beta \epsilon_i} (\beta \epsilon_i + \ln Z)$$

$$= \frac{1}{T} \sum_i \frac{e^{-\beta \epsilon_i}}{Z} \epsilon_i + \frac{k}{Z} \ln Z \sum_i e^{-\beta \epsilon_i} \tag{A.22}$$

or

$$S = \frac{U}{T} + k \ln Z \tag{A.23}$$

where

$$U \equiv \sum_i P_i \epsilon_i \tag{A.24}$$

The quantity U defined in equation (A.24) is called the "internal energy" of a system. Let us now imagine that a very small change in U is induced by an arbitrary process, which may involve interactions between the system and the outside world. We can express the fact that this infinitesimal alteration in internal energy may be due either to slight changes in the energy levels ϵ_i or to slight changes in the probabilities P_i by writing:

$$dU = \sum_i P_i d\epsilon_i + \sum_i \epsilon_i dP_i \qquad (A.25)$$

To the first term on the right-hand side of equation (A.25) we give the name "dw":

$$dw \equiv \sum_i P_i d\epsilon_i \qquad (A.26)$$

while the other term is named "dq".

$$dq \equiv dU - dw = \sum_i \epsilon_i dP_i \qquad (A.27)$$

What is the physical interpretation of these two terms? The first term, dw, involves changes in the energy levels of system, and this can only happen if we change the parameters defining the system in some way. For example, if the system is a cylinder filled with gas particles and equipped with a piston, we can push on the piston and decrease the volume available to the gas particles. This action will raise the energy levels, and when we perform it we do work on the system — work in the sense defined by Carnot, force times distance, the force which we apply to the piston multiplied by the distance through which we push it. Thus dw can be interpreted as a small amount of work performed on the system by someone or something on the outside. Another way to change the internal energy of the system is to transfer heat to it; and when a small amount of heat is transferred, the energy levels do not change, but the probabilities P_i must change slightly, as can be seen from equations (A.13), (A.14) and (A.16). Thus the quantity dq in equation (A.27) can be interpreted as an infinitesimal amount of heat transferred to the system. We have in fact anticipated this interpretation by giving it the same name as the dq of equations (4.2) and (4.3). If the probabilities P_i are changed very slightly, then from equation (20) it follows that the resulting small change in entropy is

$$dS = -k \sum_i [\ln P_i dP_i + dP_i] \qquad (A.28)$$

From equations (A.13) and (A.14) it follows that

$$\sum_i P_i = 1 \qquad (A.29)$$

as we would expect from the fact that P_i is interpreted as the probability that the system is in a particular state i. Therefore

$$\sum_i dP_i = d\sum_i P_i = 0 \qquad (A.30)$$

and as a consequence, the second term on the right-hand side of equation (4.31) vanishes. Making use of equation (A.14) to rewrite $\ln P_i$, we then have:

$$dS = -k\sum_i [(-\beta\epsilon_i - \ln Z)dP_i] \qquad (A.31)$$

or

$$dS = \frac{1}{T}\sum_i \epsilon_i dP_i = \frac{dq}{T} \qquad (A.32)$$

The somewhat complicated discussion which we have just gone through is a simplified paraphrase of Boltzmann's argument showing that if he defined entropy to be proportional to $\ln W$ (the equation engraved on his tombstone) then the function which he defined in this way must be identical with the entropy of Clausius. (We can perhaps sympathize with Ostwald and Mach, who failed to understand Boltzmann!)

Appendix B

BIOSEMIOTICS

The Oxford Dictionary of Biochemistry and Molecular Biology (Oxford University Press, 1997) defines biosemiotics as "the study of signs, of communication, and of information in living organisms". The biologists Claus Emmeche and K. Kull offer another definition of biosemiotics: "biology that interprets living systems as sign systems".

The American philosopher Charles Sanders Peirce (1839-1914) is considered to be one of the founders of semiotics (and hence also of biosemiotics). Peirce studied philosophy and chemistry at Harvard, where his father was a professor of mathematics and astronomy. He wrote extensively on philosophical subjects, and developed a theory of signs and meaning which anticipated many of the principles of modern semiotics. Peirce built his theory on a triad: (1) the sign, which represents (2) something to (3) somebody. For example, the sign might be a broken stick, which represents a trail to a hunter, it might be the arched back of a cat, which represents an aggressive attitude to another cat, it might be the waggle-dance of a honey bee, which represents the coordinates of a source of food to her hive-mates, or it might be a molecule of trans-10-cis-hexadecadienol, which represents irresistible sexual temptation to a male moth of the species Bombyx mori. The sign might be a sequence of nucleotide bases which represents an amino acid to the ribosome-transfer-RNA system, or it might be a cell-surface antigen which represents self or non-self to the immune system. In information technology, the sign might be the presence or absence of a pulse of voltage, which represents a binary digit to a computer. Semiotics draws our attention to the sign and to its function, and places much less emphasis on the physical object which forms the sign. This characteristic of the semiotic viewpoint has been expressed by the Danish biologist Jesper Hoffmeyer in the following words: "The sign, rather than the molecule, is the basic unit for studying life."

A second important founder of biosemiotics was Jakob von Uexküll (1864–1944). He was born in Estonia, and studied zoology at the University of Tartu. After graduation, he worked at the Institute of Physiology at the University of Heidelberg, and later at the Zoological Station in Naples. In 1907, he was given an honorary doctorate by Heidelberg for his studies of the physiology of muscles. Among his discoveries in this field was the first recognized instance of negative feedback in an organism.

Von Uexküll's later work was concerned with the way in which animals experience the world around them. To describe the animal's subjective perception of its environment he introduced the word Umwelt; and in 1926 he founded the Institut fur Umweltforschung at the University of Heidelberg.

Von Uexküll visualized an animal — for example a mouse — as being surrounded by a world of its own — the world conveyed by its own special senses organs, and processed by its own interpretative systems. Obviously, the Umwelt will differ greatly depending on the organism. For example, bees are able to see polarized light and ultraviolet light; electric eels are able to sense their environment through their electric organs; many insects are extraordinarily sensitive to pheromones; and a dog's Umwelt is far richer in smells than that of most other animals. The Umwelt of a jellyfish is very simple, but nevertheless it exists. Von Uexküll's Umwelt concept can even extend to one-celled organisms, which receive chemical and tactile signals from their environment, and which are often sensitive to light.

It is interesting to ask to what extent the concept of Umwelt can be equated to that of consciousness. To the extent that these two concepts can be equated, von Uexküll's Umweltforschung offers us the opportunity to explore the phylogenetic evolution of the phenomenon of consciousness.

The ideas and research of Jakob von Uexküll inspired the later work of the Nobel Laureate ethologist Konrad Lorenz, and thus von Uexküll can be thought of as one of the founders of ethology as well as of biosemiotics. Indeed, ethology and biosemiotics are closely related.

Biosemiotics also values the ideas of the American anthropologist Gregory Bateson (1904–1980), who was mentioned in Chapter 7 in connection with cybernetics and with the Macy Conferences. He was married to another celebrated anthropologist, Margaret Mead, and together they applied Norbert Wiener's insights concerning feedback mechanisms to sociology, psychology and anthropology. Bateson was the originator of a famous epigrammatic definition of information: "..a difference which makes a difference". This definition occurs in Chapter 3 of Bateson's book, Mind and

Nature: A Necessary Unity, Bantam, (1980), and its context is as follows: "To produce news of a difference, i.e., information", Bateson wrote, "there must be two entities... such that news of their difference can be represented as a difference inside some information-processing entity, such as a brain or, perhaps, a computer. There is a profound and unanswerable question about the nature of these two entities that between them generate the difference which becomes information by making a difference. Clearly each alone is — for the mind and perception — a non-entity, a non-being... the sound of one hand clapping. The stuff of sensation, then, is a pair of values of some variable, presented over time to a sense organ, whose response depends on the ratio between the members of the pair".

Suggestions for further reading

(1) J. Hoffmeyer, *Some semiotic aspects of the psycho-physical relation: the endo-exosemiotic boundary*, in Biosemiotics. The Semiotic Web, T.A. Sebeok and J. Umiker-Sebeok, editors, Mouton de Gruyter, Berlin/New York, (1991).

(2) J. Hoffmeyer, *The swarming cyberspace of the body*, Cybernetics and Human Knowing, **3(1)**, 1-10 (1995).

(3) J. Hoffmeyer, *Signs of Meaning in the Universe*, Indiana University Press, Bloomington IN, (1996).

(4) J. Hoffmeyer, *Biosemiotics: Towards a new synthesis in biology*, European J. Semiotic Stud. **9(2)**, 355-376 (1997).

(5) J. Hoffmeyer and C. Emmeche, *Code-duality and the semiotics of nature*, in On Semiotic Modeling, M. Anderson and F. Merrell, editors, Mouton de Gruyter, New York, (1991).

(6) C. Emmeche and J. Hoffmeyer, *From language to nature — The semiotic metaphor in biology*, Semiotica, **84**, 1-42 (1991).

(7) C. Emmeche, *The biosemiotics of emergent properties in a pluralist ontology*, in Semiosis, Evolution, Energy: Towards a Reconceptualization of the Sign, E. Taborsky, editor, Shaker Verlag, Aachen, (1999).

(8) S. Brier, *Information and consciousness: A critique of the mechanistic concept of information*, in Cybernetics and Human Knowing, **1(2/3)**, 71-94 (1992).

(9) S. Brier, *Ciber-Semiotics: Second-order cybernetics and the semiotics of C.S. Peirce*, Proceedings from the Second European Congress on System Science, Prague, October 5-8, 1993, AFCET, (1993).

(10) S. Brier, *A cybernetic and semiotic view on a Galilean theory of psy-*

chology, Cybernetics and Human Knowing, **2 (2)**, 31-46 (1993).

(11) S. Brier, *Cybersemiotics: A suggestion for a transdisciplinary framework for description of observing, anticipatory, and meaning producing systems*, in D.M. Dubois, editor, Computing Anticipatory Systems, CASYS — First International Conference, Liege, Belgium 1997, AIP Conference Proceedings no. 437, (1997).

(12) S. Oyama, *The Ontogeny of Information*, Cambridge University Press, (1985).

(13) J. Hoffmeyer, *The swarming cyberspace of the body*, Cybernetics and Human Knowing, **3(1)**, 1-10 (1995).

(14) J.L. Casti and A. Karlqvist, editors, *Complexity, Language, and Life: Mathematical Approaches*, Springer, Berlin, (1985).

(15) H. Maturana and F. Varla, *Autopoiesis and Cognition: The Realization of the Living*, Reidel, London, (1980).

(16) J. Mingers, *Self-Producing Systems: Implications and Application of Autopoiesis*, Plenum Press, New York, (1995).

(17) J. Buchler, editor, *Philosophical Writings of Peirce: Selected and Edited with an Introduction by Justus Buchler*, Dover Publications, New York, (1955).

(18) T.L. Short, *Peirce's semiotic theory of the self*, Semiotica, **91 (1/2)**, 109-131 (1992).

(19) J. von Uexküll, *Umwelt und Innenwelt der Tiere. 2. verm, und verb. Aufl.*, Springer, Berlin, (1921).

(20) J. von Uexküll, The theory of meaning, Semiotica, 42(1), 25-87 (1982 [1940]).

(21) T. von Uexküll, *Introduction: Meaning and science in Jacob von Uexkull's concept of biology*, Semiotica, **42**, 1-24 (1982).

(22) T. von Uexküll, *Medicine and semiotics*, Semiotica, **61** , 201-217 (1986).

(23) G. Bateson, *Form, substance, and difference. Nineteenth Annual Korzybski Memorial Lecture*, (1970). Reprinted in G. Bateson, Steps to an Ecology of Mind, Balentine Books, New York, (1972), pp. 448-464.

(24) G. Bateson, *Mind and Nature: A Necessary Unity*, Bantam Books, New York, (1980).

(25) G. Bateson, *Sacred Unity: Further Steps to an Ecology of Mind*, Harper Collins, New York, (1991).

(26) J. Ruesch and G. Bateson, *Communication*, Norton, New York, (1987).

(27) E.F. Yates, *Semiotics as a bridge between information (biology) and*

dynamics (physics), Recherches Semiotiques/Semiotic Inquiry **5**, 347-360 (1985).

(28) T.A. Sebeok, *Communication in animals and men*, Language, **39**, 448-466 (1963).

(29) T.A. Sebeok, *The Sign and its Masters*, University of Texas Press, (1979).

(30) P. Bouissac, *Ecology of semiotic space: Competition, exploitation, and the evolution of arbitrary signs*, Am. J. Semiotics, **10**, 145-166 (1972).

(31) F. Varla, *Autopoiesis: A Theory of Living Organization*, North Holland, New York, (1986).

(32) R. Posner, K. Robins and T.A. Sebeok, editors, *Semiotics: A Handbook of the Sign-Theoretic Foundations of Nature and Culture*, Walter de Gruyter, Berlin, (1992).

(33) R. Paton, *The ecologies of hereditary information*, Cybernetics and Human Knowing, **5(4)**, 31-44 (1998).

(34) T. Stonier, *Information and the Internal Structure of the Universe*, Springer, Berlin, (1990).

(35) T. Stonier, *Information and Meaning: An Evolutionary Perspective*, Springer, Berlin, (1997).

Appendix C

ENTROPY AND ECONOMICS

Human society as a superorganism, with the global economy as its digestive system

Elsewhere in this book, we have focused on human cultural evolution as a phenomenon involving cybernetic information, neglecting the fact that cultural evolution also involves vast quantities of matter and energy, and hence thermodynamic information. In this Appendix, we will try to make the picture more complete. To do so, it is useful to invoke the concept of a superorganism.

All multicellular organisms can be regarded as cooperative societies of cells. On the borderline between single-celled organisms and multicellular ones, we find blue-green algae, sponges and slime molds, where the individual cells are able to live separately, but find advantages in organizing themselves into cooperative communities. At a higher level, a nest of ants, a hive of bees or a nest of termites can be thought of a superorganism. In fact, an individual ant, an individual bee or an individual termite would find it harder to survive in isolation for a long period of time than would an individual sponge cell.

What about the amazing nests that the social insects build? Should these be regarded as a part of the social-insect superorganism, just as the shell produced by a snail is regarded as a part of the snail? In my own opinion, it is more true to regard these external structures as part of the superorganism than it would be to deny that they should be seen in this way.

Let us now turn to human "nests". A completely isolated human being would find it as difficult to survive for a long period of time as would an isolated ant or bee or termite. Therefore it seems correct to regard human society as a superorganism. In the case of humans, the analog of the social insects' nest is the enormous and complex material structure of civilization. It is, in fact, what we call the human economy. It consists of functioning factories, farms, homes, transportation links, water supplies, electrical networks, computer networks and much more. Almost all of the activities of modern humans take place through the medium of these external "exosomatic" parts of our social superorganism[3].

The economy associated with the human superorganism "eats" resources and free energy. It uses these inputs to produce local order, and finally excretes them as heat and waste. The process is closely analogous to food passing through the alimentary canal of an individual organism. The free energy and resources that are the inputs of our economy drive it just as food drives the processes of our body, but in both cases, waste products are finally excreted in a degraded form.

Almost all of the free energy that drives the human economy came originally from the sun's radiation, the exceptions being geothermal energy which originates in the decay of radioactive substances inside the earth, and tidal energy, which has its origin in the slowing of the motions of the earth-moon system. However, since the start of the Industrial Revolution, our economy has been using the solar energy stored in of fossil fuels. These fossil fuels were formed over a period of several hundred million years. We are using them during a few hundred years, i.e., at a rate approximately a million times the rate at which they were formed.

The total ultimately recoverable resources of fossil fuels amount to roughly 1260 terawatt-years of energy (1 terawatt-year $\equiv 10^{12}$ Watt-years $\equiv 1$ TWy is equivalent to 5 billion barrels of oil or 1 billion tons of coal). Of this total amount, 760 TWy is coal, while oil and natural gas each constitute roughly 250 TWy.[4] In 1890, the rate of global consumption of energy was 1 terawatt, but by 1990 this figure had grown to 13.2 TW, distributed as follows: oil, 4.6; coal, 3.2; natural gas, 2.4; hydropower, 0.8; nuclear, 0.7; fuelwood, 0.9; crop wastes, 0.4; and dung, 0.2. By 2005, the rate of oil, natural gas and coal consumption had risen to 6.0 TW, 3.7 TW and 3.5 TW respectively. Thus the present rate of consumption of fossil fuels is more than 13 terawatts and, if used at the present rate, fossil fuels would last less than a century. However, because of the very serious threats

---

[3]The terms "exosomatic" and "endosomatic" were coined by the American scientist Alfred Lotka (1820-1949). A lobster's claw is endosomatic - it is part of the lobster's body. The hammer used by a human is exosomatic - like a detachable claw. Lotka spoke of "exosomatic evolution", including in this term not only cultural evolution but also the building up of the material structures of civilization.

[4]British Petroleum, "B.P. Statistical Review of World Energy", London, 1991.

posed by climate change, human society would be well advised to stop the consumption of coal, oil and natural gas well before that time.

The rate of growth of of new renewable energy sources is increasing rapidly. These sources include small hydro, modern biomass, solar, wind, geothermal, wave and tidal energy. However, these sources currently account for only 2.8% of total energy use. There is an urgent need for governments to set high taxes on fossil fuel consumption and to shift subsidies from the petroleum and nuclear industries to renewables. These changes in economic policy are needed to make the prices of renewables more competitive.

The shock to the global economy that will be caused by the end of the fossil fuel era will be compounded by the scarcity of other non-renewable resources, such as metals. While it is true (as neoclassical economists emphasize) that "matter and energy can neither be created nor destroyed", free energy can be degraded into heat, and concentrated deposits of minerals can be dispersed. Both the degradation of Gibbs free energy into heat and the dispersal of minerals involve increases of entropy.

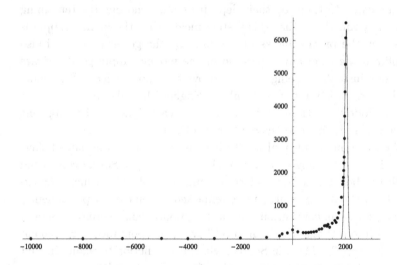

Fig. C.1 This graph shows population growth and fossil fuel use, seen on a time-scale of several thousand years. The dots are population estimates in millions from the US Census Bureau. The spike-like solid curve shows fossil fuel use, rising from almost nothing to a high value, and then falling again to almost nothing in the space of a few centuries. When the two curves are plotted together, the explosive rise of global population is seen to be simultaneous with, and perhaps partially driven by, the rise of fossil fuel use. This raises the question of whether the end of the fossil fuel era will cause a crash in the global human population.

Frederick Soddy

One of the first people to call attention to the relationship between entropy and economics was the English radiochemist Frederick Soddy (1877-1956). Soddy won the Nobel Prize for Chemistry in 1926 for his work with Ernest Rutherford demonstrating the transmutation of elements in radioactive decay processes. His concern for social problems then led him to a critical study of the assumptions of classical economics.

Soddy believed that there is a close connection between Gibbs free energy and wealth, but only a very tenuous connection between wealth and money. He was working on these problems during the period, after World War I, when England left the gold standard, and he advocated an index system to replace it. In this system, the Bank of England would print more money and lend it to private banks whenever the cost of standard items indicated that too little money was in circulation, or conversely reabsorb printed money if the index showed the money supply to be too large.

Soddy was extremely critical of the system of "fractional reserve banking" whereby private banks keep only a very small fraction of the money that is entrusted to them by their depositors and lend out the remaining amount. He pointed out that this system means that the money supply is controlled by the private banks rather than by the government, and also that profits made from any expansion of the money supply go to private corporations instead of being used to provide social services. Fractional reserve banking exists today, not only in England but also in many other countries. Soddy's criticisms of this practice cast light on the subprime mortgage crisis of 2008 and the debt crisis of 2011.

As Soddy pointed out, real wealth is subject to the second law of thermodynamics. As entropy increases, real wealth decays. Soddy contrasted this with the behavior of debt at compound interest, which increases exponentially without any limit, and he remarked: "You cannot permanently pit an absurd human convention, such as the spontaneous increment of debt [compound interest] against the natural law of the spontaneous decrement of wealth [entropy]". Thus, in Soddy's view, it is a fiction to maintain that being owed a large amount of money is a form of real wealth.

Frederick Soddy's book, *Wealth, virtual wealth and debt: The solution of the economic paradox*, published in 1926 by Allen and Unwin, was received by the professional economists of the time as the quixotic work of an outsider. Today, however, Soddy's common-sense economic analysis is increasingly valued for the light that it throws on the problems of our fractional reserve banking system, which becomes more and more vulnerable to failure as economic growth falters.

Nicholas Georgescu-Roegen

The incorporation of the idea of entropy into economic thought also owes much to the mathematician and economist Nicholas Georgescu-Roegen (1906-1994), the son a Romanian army officer. Georgescu-Roegen's talents were soon recognized by the Romanian school system, and he was given an outstanding education in mathematics, which later contributed to his success and originality as an economist.

Between 1927 and 1930 the young Georgescu studied at the Institute de Statistique in Paris, where he completed an award-winning thesis: *On the problem of finding out the cyclical components of phenomena.* He then worked in England with Karl Pearson from 1930 to 1932, and during this period his work attracted the attention of a group of economists who were working on a project called the Harvard Economic Barometer. He received a Rockefeller Fellowship to join this group, but when he arrived at Harvard, he found that the project had been disbanded. In desperation, Georgescu-Roegen asked the economist Joseph Schumpeter for an appointment to his group. Schumpeter's group was in fact a remarkably active and interesting one, which included the future Nobel laureate Wassely Leontief; and there followed a period of intense intellectual activity during which Georgescu-Roegen became an economist.

Despite offers of a permanent position at Harvard, Georgescu-Roegen returned to his native Roumania in the late 1930's and early 1940's in order to help his country. He served as a member of the Central Committee of the Romanian National Peasant Party. His experiences at this time led to his insight that economic activity involves entropy. He was also helped to this insight by Borel's monograph on Statistical Mechanics, which he had read during his Paris period.

Georgescu-Roegen later wrote: "The idea that the economic process is not a mechanical analogue, but an entropic, unidirectional transformation began to turn over in my mind long ago, as I witnessed the oil wells of the Plosti field of both World Wars' fame becoming dry one by one, and as I grew aware of the Romanian peasants' struggle against the deterioration of their farming soil by continuous use and by rains as well. However it was the new representation of a process that enabled me to crystallize my thoughts in describing the economic process as the entropic transformation of valuable natural resources (low entropy) into valueless waste (high entropy)."

After making many technical contributions to economic theory, Georgescu-Roegen returned to this insight in his important 1971 book, *The Entropy Law and the Economic Process* (Harvard University Press), where he outlines his concept of bioeconomics. In a later book, *Energy and Economic Myths* (Pergamon Press, New York, 1976), he offered the following recommendations for moving towards a bioeconomic society:

- the complete prohibition of weapons production, thereby releasing productive forces for more constructive purposes;
- immediate aid to underdeveloped countries;
- gradual decrease in population to a level that could be maintained only by organic agriculture;
- avoidance, and strict regulation if necessary, of wasteful energy use;
- abandon our attachment to "extravagant gadgetry";
- "get rid of fashion";
- make goods more durable and repairable; and
- cure ourselves of workaholic habits by rebalancing the time spent on work and leisure, a shift that will become incumbent as the effects of the other changes make themselves felt.

Georgiescu-Roegen did not believe that his idealistic recommendations would be adopted, and he feared that human society is headed for a crash.

Limits to Growth — A steady-state economy

Nicholas Georgescu-Roegen's influence continues to be felt today, not only through his own books and papers but also through those of his student, the distinguished economist Herman E. Daly, who for many years has been advocating a steady-state economy. As Daly points out in his books and papers, it is becoming increasingly apparent that unlimited economic growth on a finite planet is a logical impossibility. However, it is important to distinguish between knowledge, wisdom and culture, which can and should continue to grow, and growth in the sense of an increase in the volume of material goods produced. It is growth in the latter sense that is reaching its limits.

Daly describes our current situation as follows: "The most important change in recent times has been the growth of one subsystem of the Earth, namely the economy, relative to the total system, the ecosphere. This huge

THE LIMITS TO

Donella H. Meadows
Dennis L. Meadows
Jørgen Randers
William W. Behrens III

A Report for THE CLUB OF ROME'S Project on the
Predicament of Mankind

A POTOMAC ASSOCIATES BOOK $2.75

Fig. C.2 In 1968 Aurelio Pecci, Thorkil Kristensen and others founded the Club of
Rome, an organization of economists and scientists devoted to studying the predicament
of human society. One of the first acts of the organization was to commission an MIT
study of future trends using computer models. The result was a book entitled "Limits to
Growth", published in 1972. From the outset the book was controversial, but it became
a best-seller. It was translated into many languages and sold 30 million copies. The book
made use of an exponential index for resources, i.e., the number of years that a resource
would last if used at an exponentially increasing rate. Today the more accurate Hubbert
Peak model is used instead to predict rate of use of a scarce resource as a function of
time. Although the specific predictions of resource availability in "Limits to Growth"
lacked accuracy, its basic thesis — that unlimited economic growth on a finite planet is
impossible — was indisputably correct. Nevertheless the book was greeted with anger
and disbelief by the community of economists, and these emotions still surface when it
is mentioned.

shift from an "empty" to a "full" world is truly 'something new under the
sun'... The closer the economy approaches the scale of the whole Earth,
the more it will have to conform to the physical behavior mode of the
Earth... The remaining natural world is no longer able to provide the
sources and sinks for the metabolic throughput necessary to sustain the
existing oversized economy - much less a growing one. Economists have
focused too much on the economy's circulatory system and have neglected
to study its digestive tract."

Thorkil Kristensen, former Secretary General of the Organization for
Economic Cooperation and Development and one of the founders of the

Club of Rome, expressed the same thought in the following words: "Let us try to translate pollution and ruthless exploitation of the environment into economic language: Both of these mean that we are spending our capital, i.e., we are spending the earth's riches of coal, oil and raw materials, as well as our inheritance of clean air, clean water, and places where one can be free from noise pollution. It is clear that economic growth, as we experience it today, means that we are spending more and more of humankind's natural wealth. This cannot continue indefinitely."

Economic activity is usually divided into two categories, (1) production of goods and (2) provision of services. It is the rate of production of goods that will be limited by the carrying capacity of the global environment. Services that have no environmental impact will not be constrained in this way. Thus a smooth transition to a sustainable economy will involve a shift of a large fraction the work force from the production of goods to the provision of services.

In his recent popular book *The Rise of the Creative Class*, the economist Richard Florida points out that in a number of prosperous cities — for example Stockholm — a large fraction of the population is already engaged in what might be called creative work — a type of work that uses few resources, and produces few waste products — work which develops knowledge and culture rather than producing material goods. For example, producing computer software requires few resources and results in few waste products. Thus it is an activity with a very small ecological footprint. Similarly, education, research, music, literature and art are all activities that do not weigh heavily on the carrying capacity of the global environment. Furthermore, cultural activities lead in a natural way to global cooperation and internationalism, since cultural achievements are shared by the people of the entire world. Indeed, the shared human inheritance of culture and knowledge is growing faster than ever before. Florida sees this as a pattern for the future, and maintains that everyone is capable of creativity. He visualizes the transition to a sustainable future economy as one in which a large fraction of the work force moves from industrial jobs to information-related work. Meanwhile, as Florida acknowledges, industrial workers feel uneasy and threatened by such trends.

Biological carrying capacity and economics

Classical economists pictured the world as largely empty of human activities. According to the empty-world picture of economics, the limiting factors in the production of food and goods are shortages of human capital and labor. The land, forests, fossil fuels, minerals, oceans filled with fish, and other natural resources upon which human labor and capital operate, are assumed to be present in such large quantities that they are not limiting factors. In this picture, there is no naturally-determined upper limit to the total size of the human economy. It can continue to grow as long as new capital is accumulated, as long as new labor is provided by population growth, and as long as new technology replaces labor by automation.

Biology, on the other hand, presents us with a very different picture. Biologists remind us that if any species, including our own, makes demands on its environment which exceed the environment's carrying capacity, the result is a catastrophic collapse both of the environment and of the population which it supports. Only demands which are within the carrying capacity are sustainable. For example, there is a limit to regenerative powers of a forest. It is possible to continue to cut trees in excess of this limit, but only at the cost of a loss of forest size, and ultimately the collapse and degradation of the forest. Similarly, cattle populations may for some time exceed the carrying capacity of grasslands, but the ultimate penalty for overgrazing will be degradation or desertification of the land. Thus, in biology, the concept of the carrying capacity of an environment is extremely important; but in economic theory this concept has not yet been given the weight which it deserves.

The terminology of economics can be applied to natural resources: For example, a forest can be thought of as natural capital, and the sustainable yield from the forest as interest. Exceeding the biological carrying capacity then corresponds, in economic terms, to spending one's capital. It is easy to exceed the carrying capacity of an environment without realizing it. The populations of many species of wild animals exhibit oscillations which are produced when a population increases beyond the limits of sustainability and then crashes. It seems likely that the earth's population of humans is headed for a similar overshoot of the sustainable limits of its biophysical support system, followed by a crash.

There is much evidence indicating that the total size of the human industrial economy is very rapidly approaching the absolute limits imposed by the carrying capacity of the global environment. For example, a recent study by Vitousek et al. showed that 40 percent of the net primary product of landbased photosynthesis is appropriated, directly or indirectly, for human use. Thus, we are only a single doubling time away from 80 percent appropriation, which would certainly imply a disastrous degradation of the natural environment[5].

Another indication of our rapid approach to the absolute limit of environmental carrying capacity can be found in the present rate of loss of biodiversity. The total number of species of living organisms on the earth is thought to be between 5 million and 30 million, of which only 1.4 million have been described. Between 50 percent and 90 percent of these species live in tropical forests, a habitat which is rapidly being destroyed, because of pressures from exploding human populations. 55 percent of the earth's tropical forests have already been cleared and burned; and an additional area four times the size of Switzerland is lost every year. Because of this loss of habitat, tropical species are now becoming extinct at a rate which is many thousands of times the normal background rate.

If losses continue at the present rate, 20 percent of all tropical species will vanish irrevocably within the next 50 years. One hardly dares to think of what will happen after that. The beautiful and complex living organisms on our planet are the product of more than three billion years of evolution; but today, delicately balanced and intricately interrelated communities of living things are being destroyed on a massive scale by human greed and thoughtlessness.

Further evidence that the total size of the human economy has reached or exceeded the limits of sustainability comes from global warming, from the destruction of the ozone layer, from the rate of degradation and desertification of land, from statistics on rapidly vanishing non-renewable resources, and from recent famines. Instead of burning our tropical forests, it might be wise for us to burn our books on growth-oriented economics. Certainly an entirely new form of economics is needed today not the empty-world economics of Adam Smith, but what might be called "full-world economics", or "equilibrium economics".

Adam Smith was perfectly correct in saying that the free market is the dynamo of economic growth; but exponential growth of human population and industrial activity have brought us, in a surprisingly short time, from the empty-world situation in which he lived to a full-world situation. In today's world, we are pressing against the absolute limits of the earth's carrying capacity, and further growth carries with it the danger of future collapse. Full-world economics, the economics of the future, will no longer be able to rely on industrial growth to give profits to stockbrokers or to

---

[5] The net primary product of photosynthesis is defined as the total quantity of solar energy converted to chemical energy by plants, minus the energy used by the plants themselves.

solve problems of unemployment or to alleviate poverty. In the long run, neither growth of population nor growth of industry is sustainable; and we have now reached or exceeded the sustainable limits.

The limiting factors in economics are no longer the supply of capital or human labor or even technology. The limiting factors are the rapidly vanishing supplies of petroleum and metal ores, the forests damaged by acid rain, the diminishing catches from overfished oceans, and the cropland degraded by erosion or salination, or lost to agriculture under a cover of asphalt. Neoclassical economists have maintained that it is generally possible to substitute man-made capital for natural resources; but a closer examination shows that there are only very few cases where this is really practical.

The size of the human economy is, of course, the product of two factors the total number of humans, and the consumption per capita. If we are to achieve a sustainable global society in the future, a society whose demands are within the carrying capacity of of the global environment, then both these factors must be reduced. The responsibility for achieving sustainability is thus evenly divided between the North and the South: Where there is excessively high consumption per capita, it must be reduced; and this is primarily the responsibility of the industrialized countries. High birth rates must also be reduced; and this is primarily the responsibility of the developing countries. Both of these somewhat painful changes are necessary for sustainability; but both will be extremely difficult to achieve because of the inertia of institutions, customs and ways of thought which are deeply embedded in society, in both the North and the South.

Population and food supply

Let us look first at the problem of high birth rates: The recent spread of modern medical techniques throughout the world has caused death rates to drop sharply; but since social customs and attitudes are slow to change, birth rates have remained high. As a result, between 1930 and 2011, the population of the world increased with explosive speed from two billion to seven billion.

During the last few decades, the number of food-deficit countries has lengthened; and it now reads almost like a United Nations roster. The food-importing nations are dependent, almost exclusively, on a single food-exporting region, the grain belt of North America. In the future, this region may be vulnerable to droughts produced by global warming.

An analysis of the global ratio of population to cropland shows that we probably already have exceeded the sustainable limit of population through our dependence on petroleum: Between 1950 and 1982, the use of cheap petroleum-derived fertilizers increased by a factor of 8, and much of our present agricultural output depends their use. Furthermore, petroleum-derived synthetic fibers have reduced the amount of cropland needed for growing natural fibers, and petroleum-driven tractors have replaced draft animals which required cropland for pasturage. Also, petroleum fuels have replaced fuelwood and other fuels derived for biomass. The reverse transition, from fossil fuels back to renewable energy sources, will require a considerable diversion of land from food production to energy production.

As population increases, the cropland per person will continue to fall, and we will be forced to make still heavier use of fertilizers to increase output per hectare. Also marginal land will be used in agriculture, with the probable result that much land will be degraded through erosion or salination. Reserves of oil are likely to be exhausted by the middle of this century. Thus there is a danger that just as global population reaches the unprecedented level of 9 billion or more, the agricultural base for supporting it may suddenly collapse. The resulting ecological catastrophe, possibly compounded by war and other disorders, could produce famine and death on a scale unprecedented in history - a catastrophe of unimaginable proportions, involving billions rather than millions of people. The present tragic famine in Africa is to this possible future disaster what Hiroshima is to the threat of thermonuclear war a tragedy of smaller scale, whose horrors should be sufficient, if we are wise, to make us take steps to avoid the larger catastrophe.

At present a child dies from starvation every six seconds — five million children die from hunger every year. Over a billion people in today's world are chronically undernourished. There is a threat that unless prompt and well-informed action is taken by the international community, the tragic loss of life that is already being experienced will increase to unimaginable proportions.

As glaciers melt in the Himalayas, threatening the summer water supplies of India and China; as ocean levels rise, drowning the fertile rice-growing river deltas of Asia; as aridity begins to decrease the harvests of Africa, North America and Europe; as populations grow; as aquifers are overdrawn; as cropland is lost to desertification and urban growth; and as energy prices increase, the billion people who now are undernourished but still survive, might not survive. They might become the victims of a famine whose proportions could exceed anything that the world has previously experienced.

It is vital for the world to stabilize its population, not only because of the threat of a catastrophic future famine, but also because rapid population growth is closely linked with poverty. Today, a large fraction of the world's people live in near-poverty or absolute poverty, lacking safe water, sanitation, elementary education, primary health care and proper nutrition. Governments struggling to solve these problems, and to provide roads, schools, jobs and medical help for all their citizens, find themselves defeated by the rapid doubling times of populations. For example, in Liberia, the rate of population growth is 4 percent per year, which means that the population of Liberia doubles in size every eighteen years. Under such circumstances, in spite of the most ambitious development programs, the infrastructure per capita decreases. Also, since new jobs must be found for the new millions added to the population, the introduction of efficient modern methods in industry and agriculture aggravates the already-serious problem of unemployment.

Education of women and higher status for women are vitally important measures, not only for their own sake, but also because in many countries these social reforms have proved to be strongly correlated with lower birth rates. Religious leaders who oppose programs for the education of women and for family planning on "ethical" grounds should think carefully about the scope and consequences of the catastrophic global famine which will undoubtedly occur within the next 50 years if population is allowed to increase unchecked.

Sir Partha Dasgupta of Cambridge University has pointed out that the changes needed to break the cycle of overpopulation and poverty are all desirable in themselves. Besides education and higher status for women, these measures include state-provided social security for old people, provision of water supplies near to dwellings, provision of health services to all, abolition of child labor and general economic development.

Social values and levels of consumption

Let us next turn to the problem of reducing the per-capita consumption in the industrialized countries. The whole structure of western society seems designed to push its citizens in the opposite direction, towards ever-increasing levels of consumption. The mass media hold before us continually the ideal of a personal utopia filled with material goods.

Every young man in a modern industrial society feels that he is a failure unless he fights his way to the "top"; and in recent years, women too have been drawn into this competition. Of course not everyone can reach the top; there would not be room for everyone; but society urges all us to try, and we feel a sense of failure if we do not reach the goal. Thus, modern life has become a struggle of all against all for power and possessions.

One of the central problems in reducing consumption is that in our present economic and social theory, consumption has no upper bound; there is no definition of what is enough; there is no concept of a state where all of the real needs of a person have been satisfied. In our growth-oriented present-day economics, it is assumed that, no matter how much a person earns, he or she is always driven by a desire for more.

The phrase "conspicuous consumption" was invented by the Norwegian-American economist Thorstein Veblen (1857–1929) in order to describe the way in which our society uses economic waste as a symbol of social status. In *The Theory of the Leisure Class*, first published in 1899, Veblen pointed out that it wrong to believe that human economic behavior is rational, or that it can be understood in terms of classical economic theory. To understand it, Veblen maintained, one might better make use of insights gained from anthropology, psychology, sociology, and history.

The sensation caused by the publication of Veblen's book, and the fact that his phrase, "conspicuous consumption", has become part of our language, indicate that his theory did not completely miss its mark. In fact, modern advertisers seem to be following Veblen's advice: Realizing that much of the output of our economy will be used for the purpose of establishing the social status of consumers, advertising agencies hire psychologists to appeal to the consumer's longing for a higher social position.

When possessions are used for the purpose of social competition, demand has no natural upper limit; it is then limited only by the size of the human ego, which, as we know, is boundless. This would be all to the good if unlimited economic growth were desirable. But today, when further growth implies future collapse, industrial society urgently needs to find new values to replace our worship of power, our restless chase after excitement, and our admiration of excessive consumption.

The values which we need, both to protect nature from civilization and to protect civilization from itself, are perhaps not new: Perhaps it would be more correct to say that we need to rediscover ethical values which once were part of human culture, but which were lost during the process of industrialization when technology allowed us to break traditional environmental constraints.

Our ancestors were hunter-gatherers, living in close contact with nature, and respecting the laws and limitations of nature. There are many hunter-gatherer cultures existing today, from whose values and outlook we could learn much.[6] In some parts of Africa, before cutting down a tree, a man will offer a prayer of apology to the spirit of the tree, explaining why necessity has driven him to such an act. The attitude involved in this ritual is something which industrialized society needs to learn, or relearn.

Older cultures have much to teach industrial society because they already have experience with full-world situation which we are fast approaching. In a traditional culture, where change is extremely slow, population has an opportunity to expand to the limits which the traditional way of life allows, so that it reaches an equilibrium with the environment. For example, in a hunter-gatherer culture, population has expanded to the limits which can be supported without the introduction of agriculture. The density of population is, of course, extremely low, but nevertheless it is pressing against the limits of sustainability. Overhunting or overfishing would endanger the future. Respect for the environment is thus necessary for the survival of such a culture.

Similarly, in a stable, traditional agricultural society which has reached an equilibrium with its environment, population is pressing against the limits of sustainability. In such a culture, one can usually find expressed as a strong ethical principle the rule that the land must not be degraded, but must be left fertile for the use of future generations.

It would be wise for the industrialized countries to learn from the values of older traditional cultures; but what usually happens is the reverse: The unsustainable, power-worshiping, consumption-oriented values of western society are so strongly propagandized by television, films and advertising, that they overpower and sweep aside the wisdom of older societies. Today, the whole world seems to be adopting values, fashions, and standards of behavior presented in the mass media of western society. This is unfortunate, since besides showing us unsustainable levels of affluence and economic waste, the western mass media depict values and behavior patterns which are hardly worthy of imitation.

---

[6] Unfortunately, instead of learning from them, we often move in with our bulldozers and make it impossible for their way of life to continue. During the past several decades, for example, approximately one tribe of South American forest Indians has died out every year. Of the 6000 human languages now spoken, it is estimated that half will vanish during the next 50 years.

The responsibility of governments

Like a speeding bus headed for a brick wall, the earth's rapidly-growing population of humans and its rapidly-growing economic activity are headed for a collision with a very solid barrier - the carrying capacity of the global environment. As in the case of the bus and the wall, the correct response to the situation is to apply the brakes in good time, but fear prevents us from doing this. What will happen if we slow down very suddenly? Will not many of the passengers be injured? Undoubtedly. But what will happen if we hit the wall at full speed? Perhaps it would be wise, after all, to apply the brakes!

The memory of the great depression of 1929 makes us fear the consequences of an economic slowdown, especially since unemployment is already a serious problem. Although the history of the 1929 depression is frightening, it may nevertheless be useful to look at the measures which were used then to bring the global economy back to its feet. A similar level of governmental responsibility may help us during the next few decades to avoid some of the more painful consequences of the necessary transition from the economics of growth to the economics of equilibrium.

The Worldwatch Institute, Washington D.C., lists the following steps as necessary for the transition to sustainability: (1) Stabilizing population; (2) Shifting to renewable energy; (3) Increasing energy efficiency; (4) Recycling resources; (5) Reforestation and (6) Soil Conservation. All of these steps are labor-intensive; and thus, wholehearted governmental commitment to the transition to sustainability can help to solve the problem of unemployment.

In much the same way that Keynes urged Roosevelt to use governmental fiscal and financial policy to achieve social goals, we can now urge our governments to use their control of taxation to promote sustainability. For example, a slight increase in the taxes on fossil fuels could make a number of renewable energy technologies economically competitive; and higher taxes on motor fuels would be especially useful in promoting the necessary transition from private automobiles to bicycles and public transport.

The economic recession that began with the US subprime mortgage crisis of 2007 and 2008 can be seen as an opportunity. It is thought to be temporary, but it is a valuable warning of irreversible long-term changes that will come later in the 21st century when the absolute limits of industrial growth are reached. Already today we are faced with the problems of preventing unemployment and simultaneously building the infrastructure of an ecologically sustainable society.

Today's economists believe that growth is required for economic health; but at some point during this century, industrial growth will no longer be possible. If no changes have been made in our economic system when this happens, we will be faced with massive unemployment. Three changes are needed to prevent this:

(1) Labor must be moved to tasks related to ecological sustainability. These include development of renewable energy, reforestation, soil and water conservation, replacement of private transportation by public transport, and agricultural development. Health and family planning services must also be made available to all.

(2) Opportunities for employment must be shared among those in need of work, even if this means reducing the number of hours that each person works each week and simultaneously reducing the use of luxury goods, unnecessary travel, and all forms of conspicuous consumption. It will be necessary for governments to introduce laws reducing the length of the working week, thus ensuring that opportunities for employment are shared equally.

(3) The world's fractional reserve banking system needs to be reformed along the lines suggested by Frederick Soddy.

We have the chance, already today, to make these changes in our economic system. The completely unregulated free market alone has proved to be inadequate in a situation where economic growth has slowed or halted, as is very apparent in the context of the present financial crisis. But halfway through the 21st century, industrial growth will be halted permanently by ecological constraints and vanishing resources. We must construct a steady-state economic system — one that can function without industrial growth. Our new economic system needs to have a social and ecological conscience, it needs to be responsible, and it needs to have a farsighted global ethic. We have the opportunity to anticipate and prevent future shocks by working today to build a new economic system.

The introduction of ecologically or socially motivated taxes by one country may make it less able to compete with other countries that do not include externalities in their pricing. Until such reforms become universal, free trade may give unfair advantages to countries which give the least attention to social and environmental ethics. Thus free trade and globalization will become fair and beneficial only when ethical economic practices become universal.

Governments already recognize their responsibility for education. In the future, they must also recognize their responsibility for helping young people to make a smooth transition from education to secure jobs. If jobs are scarce, work must be shared with a spirit of solidarity among those seeking employment; hours of work (and if necessary, living standards) must be reduced to insure that all who wish it may have jobs. Market forces alone cannot achieve this. The powers of government are needed.

Governments must recognize their responsibility for thinking not only of the immediate future but also of the distant future, and their responsibility for guiding us from the insecure and socially unjust world of today to a safer and happier future world. In the world as it is today, 1.6 trillion dollars are wasted on armaments each year; and while this is going on, children in the developing countries sift through garbage dumps searching for scraps of food. In today's world, the competition for jobs and for material possessions makes part of the population of the industrial countries work so hard that they damage their health and neglect their families; and while this is going on, another part of the population suffers from unemployment, becoming vulnerable to depression, mental illness, alcoholism, drug abuse and crime. In the world of the future, which we now must build, the institution of war will be abolished, and the enormous resources now wasted on war will be used constructively. In the future world as it can be if we work to make it so, a stable population of moderate size will live without waste or luxury, but in comfort and security, free from the fear of hunger or unemployment. The world which we want will be a world of changed values, where human qualities will be valued more than material possessions. Let us try to combine wisdom and ethics from humanity's past with today's technology to build a sustainable, livable and equitable future world.

References

(1) F. Soddy, *Wealth, virtual wealth and debt;: The solution of the economic paradox*, Allen and Unwin, (1926).

(2) F. Soddy, *The Role of Money*, George Routledge & Sons Ltd, (1934). (Internet Archive Gutenberg)

(3) N. Georgescu-Roegen, *The Entropy Law and the Economic Process*, Harvard University Press, (1971).

(4) Rowbotham, M., *The Grip of Death: A Study of Modern Money, Debt Slavery and Destructive Economics*, Jon Carpenter Publishing, (1998).

(5) H.E. Daly and J. Farley, *Ecological Economics: Principles and Appli-*

cations, Island Press, (2004).

(6) H.E. Daly, *Beyond Growth: The Economics of Sustainable Development*, Beacon Press, (1997).

(7) H.E. Daly, *Valuing the Earth: Economics, Ecology, Ethics*, The MIT Press; 2nd edition, (1993).

(8) H.E. Daly and J.B. Cobb, Jr., *For The Common Good: Redirecting the Economy toward Community, the Environment, and a Sustainable Future*, Beacon Press; 2nd,Updated edition (1994).

(9) R. Goodland, H.E. Daly and S. El Serafy, *Population, Technology, and Lifestyle: The Transition To Sustainability*, Island Press, (1992).

(10) R. Florida, *The Rise of the Creative Class*, Basic Books, (2002).

(11) R. Heinberg. *The End of Growth*, New Society Publishers, Gabriola Island BC Canada, (2011).

(12) R. Goodland, H. Daly, S. El Serafy and B. von Droste, editors, *Environmentally Sustainable Economic Development: Building on Brundtland*, UNESCO, Paris, (1991).

(13) D. Meadows, D. Meadows and J. Randers, *Beyond the Limits*, Chelsea Green Publishing Co., Vermont, USA, (1992).

(14) P.M. Vitousek, P.R. Ehrlich, A.H. Ehrlich and P.A. Matson, *Human Appropriation of the Products of Photosynthesis*, Bioscience, **34**, 368-373, (1986).

(15) World Resources Institute (WRI), *Global Biodiversity Strategy*, The World Conservation Union (IUCN), United Nations Environment Programme (UNEP), (1992).

(16) E.O. Wilson, editor, *Biodiversity*, National Academy Press, Washington D.C., (1988).

(17) J. van Klinken, *Het Dierde Punte*, Uitgiversmaatschappij J.H. Kok-Kampen, Netherlands (1989).

(18) A.H. Ehrlich and U. Lele, *Humankind at the Crossroads: Building a Sustainable Food System*, in Draft Report of the Pugwash Study Group: The World at the Crossroads, Berlin, (1992).

(19) L.R. Brown, *The Twenty-Ninth Day*, W.W. Norton, (1978).

(20) L.R. Brown, *Building a Sustainable Society*, W.W. Norton, (1981).

(21) L.R. Brown and J.L. Jacobson, *Our Demographically Divided World*, Worldwatch Paper 74, Worldwatch Institute, Washington D.C., (1986).

(22) P.B. Smith, J.D. Schilling and A.P. Haines, *Introduction and Summary*, in Draft Report of the Pugwash Study Group: The World at the Crossroads, Berlin, (1992).

(23) T. Veblen, *The Theory of the Leisure Class*, Modern Library, (1934).

(24) L.R. Brown and P. Shaw, *Six Steps Towards a Sustainable Society*, Worldwatch Paper 48, Worldwatch Institute, Washington D.C., (1982).

(25) L.R. Brown, *Who Will Feed China?*, W.W. Norton, New York, (1995).

(26) L.R. Brown, et al., *Saving the Planet. How to Shape and Environmentally Sustainable Global Economy*, W.W. Norton, New York, (1991).

(27) L.R. Brown, *Postmodern Malthus: Are There Too Many of Us to Survive?*, The Washington Post, July 18, (1993).

(28) L.R. Brown and H. Kane, *Full House. Reassessing the Earth's Population Carrying Capacity*, W.W. Norton, New York, (1991).

(29) L.R. Brown, *Seeds of Change*, Praeger Publishers, New York, (1970).

(30) L.R. Brown, *The Worldwide Loss of Cropland*, Worldwatch Paper 24, Worldwatch Institute, Washington, D.C., (1978).

(31) L.R. Brown, and J.L. Jacobson, *Our Demographically Divided World*, Worldwatch Paper 74, Worldwatch Institute, Washington D.C., (1986).

(32) L.R. Brown, and J.L. Jacobson, *The Future of Urbanization: Facing the Ecological and Economic Constraints*, Worldwatch Paper 77, Worldwatch Institute, Washington D.C., (1987).

(33) L.R. Brown, and others, *State of the World*, W.W. Norton, New York, (published annually).

(34) H. Brown, *The Human Future Revisited. The World Predicament and Possible Solutions*, W.W. Norton, New York, (1978).

(35) H. Hanson, N.E. Borlaug and N.E. Anderson, *Wheat in the Third World*, Westview Press, Boulder, Colorado, (1982).

(36) A. Dil, ed., *Norman Borlaug and World Hunger*, Bookservice International, San Diego/Islamabad/Lahore, (1997).

(37) N.E. Borlaug, *The Green Revolution Revisited and the Road Ahead*, Norwegian Nobel Institute, Oslo, Norway, (2000).

(38) N.E. Borlaug, *Ending World Hunger. The Promise of Biotechnology and the Threat of Antiscience Zealotry*, Plant Physiology, **124**, 487-490, (2000).

(39) M. Giampietro and D. Pimentel, *The Tightening Conflict: Population, Energy Use and the Ecology of Agriculture*, in *Negative Population Forum*, L. Grant ed., Negative Population Growth, Inc., Teaneck, N.J., (1993).

(40) H.W. Kendall and D. Pimentel, *Constraints on the Expansion of the Global Food Supply*, Ambio, **23**, 198-2005, (1994).

(41) D. Pimentel et al., *Natural Resources and Optimum Human Population*, Population and Environment, **15**, 347-369, (1994).

(42) D. Pimentel et al., *Environmental and Economic Costs of Soil Erosion*

and *Conservation Benefits*, Science, **267**, 1117-1123, (1995).

(43) D. Pimentel et al., *Natural Resources and Optimum Human Population*, Population and Environment, **15**, 347-369, (1994).

(44) D. Pimentel and M. Pimentel, *Food Energy and Society*, University Press of Colorado, Niwot, Colorado, (1996).

(45) D. Pimentel et al., *Environmental and Economic Costs of Soil Erosion and Conservation Benefits*, Science, **267**, 1117-1123, (1995).

(46) RS and NAS, *The Royal Society and the National Academy of Sciences on Population Growth and Sustainability*, Population and Development Review, **18**, 375-378, (1992).

(47) A.M. Altieri, *Agroecology: The Science of Sustainable Agriculture*, Westview Press, Boulder, Colorado, (1995).

(48) G. Conway, *The Doubly Green Revolution*, Cornell University Press, (1997).

(49) J. Dreze and A. Sen, *Hunger and Public Action*, Oxford University Press, (1991).

(50) T. Berry, *The Dream of the Earth*, Sierra Club Books, San Francisco, (1988).

(51) G. Bridger, and M. de Soissons, *Famine in Retreat?*, Dent, London, (1970).

(52) W. Brandt, *World Armament and World Hunger: A Call for Action*, Victor Gollanz Ltd., London, (1982).

(53) A.K.M.A. Chowdhury and L.C. Chen, *The Dynamics of Contemporary Famine*, Ford Foundation, Dacca, Pakistan, (1977)

(54) J. Shepard, *The Politics of Starvation*, Carnegie Endowment for International Peace, Washington D.C., (1975).

(55) M.E. Clark, *Ariadne's Thread: The Search for New Modes of Thinking*, St. Martin's Press, New York, (1989).

(56) J.-C. Chesnais, *The Demographic Transition*, Oxford, (1992).

(57) C.M. Cipola, *The Economic History of World Population*, Penguin Books Ltd., (1974).

(58) E. Draper, *Birth Control in the Modern World*, Penguin Books, Ltd., (1972).

(59) Draper Fund Report No. 15, *Towards Smaller Families: The Crucial Role of the Private Sector*, Population Crisis Committee, 1120 Nineteenth Street, N.W., Washington D.C. 20036, (1986).

(60) E. Eckholm, *Losing Ground: Environmental Stress and World Food Prospects*, W.W. Norton, New York, (1975).

(61) E. Havemann, *Birth Control*, Time-Life Books, (1967).

(62) J. Jacobsen, *Promoting Population Stabilization: Incentives for Small Families*, Worldwatch Paper 54, Worldwatch Institute, Washington D.C., (1983).

(63) N. Keyfitz, *Applied Mathematical Demography*, Wiley, New York, (1977).

(64) W. Latz (ed.), *Future Demographic Trends*, Academic Press, New York, (1979).

(65) World Bank, *Poverty and Hunger: Issues and Options for Food Security in Developing Countries*, Washington D.C., (1986).

(66) J.E. Cohen, *How Many People Can the Earth Support?*, W.W. Norton, New York, (1995).

(67) D.W. Pearce and R.K. Turner, *Economics of Natural Resources and the Environment*, Johns Hopkins University Press, Baltimore, (1990).

(68) P. Bartelmus, *Environment, Growth and Development: The Concepts and Strategies of Sustainability*, Routledge, New York, (1994).

(69) J. Amos, *Climate Food Crisis to Deepen*, BBC News (5 September, 2005).

(70) J. Vidal and T. Ratford, *One in Six Countries Facing Food Shortage*, The Guardian, (30 June, 2005).

(71) J. Mann, *Biting the Environment that Feeds Us*, The Washington Post, July 29, 1994.

(72) G.R. Lucas, Jr., and T.W. Ogletree, (editors), *Lifeboat Ethics. The Moral Dilemmas of World Hunger*, Harper and Row, New York.

(73) J.L. Jacobson, *Gender Bias: Roadblock to Sustainable Development*, Worldwatch Paper 110, Worldwatch Institute, Washington D.C., (1992).

(74) J. Gever, R. Kaufmann, D. Skole and C. Vorosmarty, *Beyond Oil: The Threat to Food and Fuel in the Coming Decades*, Ballinger, Cambridge MA, (1986).

(75) M. ul Haq, *The Poverty Curtain: Choices for the Third World*, Columbia University Pres, New York, (1976).

(76) H. Le Bras, *La Planète au Village*, Datar, Paris, (1993).

(77) E. Mayr, *Population, Species and Evolution*, Harvard University Press, Cambridge, (1970).

Index